1985

MEDIA MANUALS

The Use of Microphones
Second Edition

MEDIA MANUALS

The
Use of
Microphones
Second Edition

Alec
Nisbett

FOCAL PRESS
London & Boston

Focal Press
is an imprint of the Butterworth Group
which has principal offices in
London, Boston, Durban, Singapore, Sydney, Toronto, Wellington

First published 1974
  Reprinted 1977, 1979
Second edition 1983
  Reprinted 1984

British Library Cataloguing in Publication Data
Nisbett, Alec
  The use of microphones.
  1. Microphone
  I. Title
  621.380′28′2   TK6478
  ISBN 0-240-51199-9

Printed and bound by Thomson Litho Ltd., East Kilbride, Scotland

Contents

621.380283
R726
2ed

$13.63

Robin Taylor

2-26-85

114, 124

Sound is a medium for communication.
High-quality sound serves communication; it is not an end in itself.

Introduction: On quality

Your end product is sound from a loudspeaker – or, occasionally, headphones. In the home it may be high-quality sound (hi-fi); more often it is simply appalling by the standards of the sound enthusiast. The immediate source may be radio, record-player, cassette or open-reel tape deck; or it may accompany pictures on TV or film. The range of possible listening conditions is wide – and we may assume that except in your own home it will generally be out of your control.

But what you can control is the quality of sound that you make available to the listener. Your first responsibility will often be towards the vast majority whose needs are simple: intelligibility and clarity of speech or a concentration on the less demanding rhythmic or melodic qualities of music. Your second responsibility will be towards those who enjoy the wider ranges of frequency and dynamics, and the greater complexity of tonal structure that may be appreciated on good equipment. This aspect of your work will give the greatest satisfaction but it is not your sole task, and because so much of this book will be concerned with how to get the best out of your equipment it is well to remember that what I am talking about is generally a best that includes the good; and that I am still very much concerned with the interests of the majority.

Sound is the medium not the message

It is also well to remember that sound quality comes second to sound content: that what is being said is generally – though not always – more important than how you say it. The sound man should do his bit to ensure that the message transcends the medium.

I am fortunate in the background that allows me both to say this and to practise it: the British Broadcasting Corporation is the biggest of its kind in the world whose contract is not with government or advertisers but directly with the audience. The range of activity that is possible in so vast an organisation, together with the simplicity of its responsibilities, makes the BBC unique both as a place to learn and as a place to do the job. This book reflects the fruits of my own experience together with that of many colleagues who have been generous with both time and advice.

Microphones and audio engineering

The microphone is the central and crucial item of sound studio equipment. It converts the mechanical energy of sound in air into electrical energy. The original sound consists of a series of variations in air pressure: the microphone turns them into a similar (but not necessarily identical) series of fluctuations in voltage. They may then be frozen in the mechanical form of a groove on a record, printed as a pattern in magnetic particles on tape, or coded piggyback on to electromagnetic waves and transmitted at the speed of light. And at the end of all this it must be possible to feed the electrical signal to a loudspeaker, which drives the air in a series of variations in pressure which closely mimic the original sound.

The judgment of sound

Reproduced sound may be similar to the original sound but it cannot be identical. In monophonic sound (mono) it comes to the listener through a single loudspeaker; in stereo from two; in quadraphonic reproduction from four, and so on. None of these can recreate the original sound field that existed in the studio; the sound is necessarily modified in some way. The results of the changes that inevitably take place can only be judged aesthetically, by ear.

The exercise of this judgment is a craft or, at best, it might even be suggested, an art. It is not a science or an aspect of engineering, although the job is often, for convenience, combined with the engineering side of sound control.

Microphone balance

The subject of this book is microphone balance: the placing of the one or many microphones which sample the sound field in the studio, the control of their output, and the aural evaluation of the results. It is primarily concerned with mono techniques, though much that is described here will also be useful in stereo. Further microphone techniques that are used only in stereo are described in *The Technique of the Sound Studio*. (See 'Further Reading', page 155.)

The microphone is treated as part of a system which includes the sound field in the studio, characteristics of microphones themselves and the equipment that is used to mix and control the resulting signals. All of these are discussed in the following pages.

An important objective of audio engineering is to minimise change, to avoid degradation of quality below some chosen level; while the sound balancing techniques that are described in this book actually promote change within certain conventions. The two functions of the sound man should not be confused.

THE STUDIO CHAIN

The *sound source* radiates sound as *pressure waves* in the air.

Some of the sound travels direct to the microphone. Some is reflected and arrives later as *reverberation.*

Unwanted sound, *noise,* also travels to the microphone.

The *microphone* converts the variations in air pressure into electrical variations – the *signal.* Imperfections in this process introduce *distortion.* The microphone and all other components also produce unwanted electronic *noise* – hiss, hum, etc.

Amplifiers may be introduced at any stage to compensate for loss in *volume* of the electrical signal.

Equalisers may be introduced at any stage to compensate for deficiencies in preceding or subsequent components (*frequency distortion*).

Artificial reverberation ('*echo*'), may be added to individual components or to the mixed output.

In a *mixer,* electrical signals from microphones, tape reproducers and record players are combined.

A *volume* or *peak programme meter* is used to ensure that the electrical level is within specified limits.

Headphones are used only when no suitable listening room is available (e.g. on film locations).

The resulting *sound balance* is judged by ear and *monitoring loudspeaker* in the listening room.

Studio output may go to a *tape recorder*; to a *transmitter* (radio or TV); or to a *closed circuit* line.

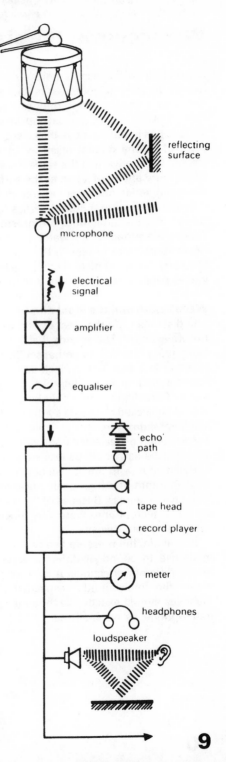

reflecting surface

microphone

electrical signal

amplifier

equaliser

'echo' path

tape head

record player

meter

headphones

loudspeaker

9

The sounds that our ear – or a microphone – hears are simply rapid fluctuations in air pressure.

Wavelength: the size of sound

A vibrating object such as a drum skin produces pressure variations: as it moves forward it compresses the air in front of it. This increase in pressure is transmitted to the next layer of air, and so on, travelling at a rate of about 1120ft per second in warm air. Meanwhile the vibrating surface has begun to move backward, creating a region of low pressure in front of it. The natural elasticity of the air is such that this in turn is transmitted outward at the same speed, to be followed by a further wave of pressure and rarefaction as the surface moves forward and back again. A regular series of waves radiates from the source.

These waves are a little different from the waves on the surface of a pond – but the similarities are greater than the differences. If you put a cork on the surface of the water it bobs up and down as each wave passes, but it stays more or less in the same place from one wave to the next. Similarly an air particle dances backward and forward as the pressure waves pass, oscillating about a median position.

Wavelength and the ideal diaphragm

The distance from the crest of one wave to the crest of the next is called the wavelength. The sounds that we can hear have wavelengths ranging from about one inch to perhaps 40ft. The orifice and diaphragm of an ear is rather less than half an inch across, about half the size of the shortest wavelength that we can hear. This is no coincidence. If it were larger some parts of the diaphragm would be subject to pressure while other regions would be sucked outwards by the rarefaction: the two would cancel out. If the diaphragm were smaller it would still register pressure and rarefaction in turn, but the area acted upon would be smaller, the total pressure would be less: the ear would be less sensitive.

By analogy with the ear we now have an ideal specification for the size of diaphragm of a high-quality pressure microphone. It should be about half an inch across. Because of the difficulty of obtaining enough sensitivity early microphones were very much bigger, but modern microphones are often about this size.

In real life, the pressure at any point fluctuates in a complex manner in response to sound of different wavelengths and intensities and from many different directions. A microphone which measures only pressure hears the total sound irrespective of the direction it comes from. It behaves almost exactly like th human ear. But this is not the only type of microphone, as we shall see.

SOUND WAVES
A vibrating panel generates waves of
pressure and rarefaction: sound. Air is
displaced backward and forward
along the line of travel.

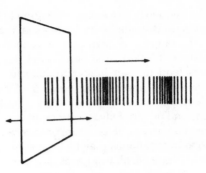

LATERAL WAVE
Diagrammatically, it is
convenient to represent the
pressure waves of sound as
lateral waves, as though the
air were displaced from side
to side as the wave passes.
In fact, this is a graph in
which displacement of air
from its median position
(vertical axis) is plotted
against time (horizontal
axis).

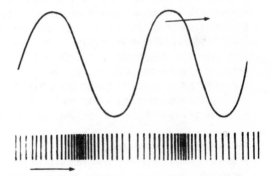

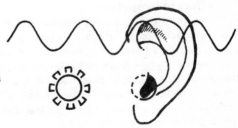

THE SHORTEST WAVELENGTH
that young ears can hear defines
the size of diaphragm that is
needed for a high-quality
pressure microphone.

11

The rate at which a surface – or an air particle – moves backward and forward is called the frequency.

Frequency: the timing of sound

The frequency of a sound measured in hertz (Hz), is the number of regular excursions made by an air particle in one second.

The range of human hearing is sometimes quoted as being from about 16Hz to 16 000Hz, the upper limit depending on age and health. Young ears tend to hear higher frequencies, but 16 000Hz is often taken as a reasonable upper limit required of engineering systems. Even so, the diaphragm of a full-range high-quality microphone must be capable of responding faithfully to very rapid changes of pressure. This means that it must be very light, or, more precisely, that it must have a very low inertia. Such microphones may be both expensive and delicate, so they are used only where absolutely necessary, e.g. for the sound o orcr orfor individual instruments like cymbals and triangles which have a strong high-frequency content. For many combinations including the human voice, a frequency response up to 10 000Hz is satisfactory.

At the other end of the scale the lowest frequencies that can be heard as musical notes merge into those that are perceived as separate puffs of air pressure. We shall see later that a perfect low-frequency response is not essential: even for the fullest range of music a frequency response that is level to about 50Hz is all that is needed.

In practice, electronic components can easily go well beyond these limits, while mechanical components may be subject to radically increased cost or reduced durability if they have the same full range. The available audio range may also be limited by external factors such as the capacity of electromagnetic or line transmissions.

Frequency and wavelength
Frequency and wavelength are closely related. The speed of sundn air can, for rough calculations, be treated as a constant. A rapidly vibrating source (high frequency) produces sound of short wavelength; a source vibrating relatively slowly (and a hundred excursions a second is slow enough to be called low frequency) produces long wavelengths.

Large objects are necessary to radiate low-frequency sound efficiently. The sound-board of a piano is suitable, but not so efficient as the larger pipes in the lower register of an organ.

The effect of temperature
A factor that does affect the speed of sound is temperature. The speed rises gradually as temperature goes up. The strings of violins can be tuned, but the vibrating column of air of most wind instruments cannot. So, when air temperature goes up and wavelength remains the same, the frequency that we hear also rises. A flute, for example, sharpens by a semitone as the temperature goes up by 15°F. This is a daily problem for orchestral musicians: they tune (to the oboe) when they start and again when the instruments are warm.

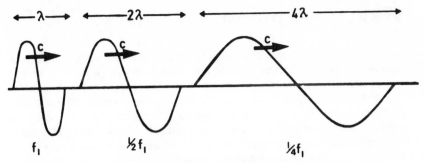

FREQUENCY AND WAVELENGTH   All sound waves travel at the same speed, c, through the same medium, so frequency, f, is inversely proportional to wavelength, λ.

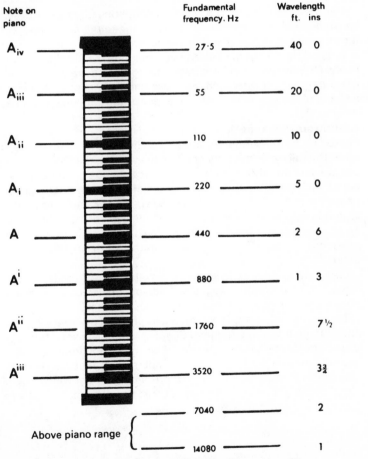

Note on piano	Fundamental frequency, Hz	Wavelength ft. ins
A_{iv}	27·5	40 0
A_{iii}	55	20 0
A_{ii}	110	10 0
A_i	220	5 0
A	440	2 6
A^i	880	1 3
A^{ii}	1760	7½
A^{iii}	3520	3¾
	7040	2
Above piano range {	14080	1

MUSICAL NOTES: FREQUENCY AND WAVELENGTH   Some examples. Frequency multiplied by wavelength equals the speed of sound, which is 1100ft/sec in cold air. (The speed is a little faster in warm air.)

13

Waves and phase

Particle velocity is the rate of movement of individual air particles. It is proportional to air pressure: where one has a peak in its waveform so has the other.

Pressure gradient is the rate at which pressure changes with distance along the wave. Where pressure is at a peak its rate of change is zero, and where pressure is that of the normal atmosphere, its rate of change is maximum. The pressure gradient waveform is similar to that for pressure, but lags a quarter of wavelength behind it.

Particle displacement is the distance that a particle of air has moved from its equilibrium position. Displacement is proportional to pressure gradient.

The mode of operation of some microphones has been described as *constant-velocity,* and others as *constant-amplitude.* These confusing terms actually mean that the output voltage is equal to a constant multiplied by diaphragm velocity or amplitude (i.e. displacement) respectively. Of more practical importance are the terms *pressure operation* and *pressure-gradient operation,* as these characteristics of a microphone's action will lead to important differences in the way that it can be used.

Adding sounds together

Phase is a term used in describing subdivisions of one wavelength of a tone. The full cycle from any point on the wave to the corresponding point one wavelength further on is a 360-degree change of phase. Here the two points are in phase; they are fully out of phase (or '180 degrees out of phase') with the point on the curve half way in between. The mathematical jargon is not important to the microphone user, but the concept of waves being in or out of phase is vital. Signals that are in phase reinforce each other; those that are out of phase subtract from or tend to cancel each other.

Complex sounds with complex waveforms are made up by adding many simple waveforms together – adding, that is, those parts that are in phase, and subtracting those that are out of phase.

The ear is not as a rule interested in phase. Two sound-waves can be added to give a whole range of different composite waveforms which all sound the same to the ear. This has important implications for microphone design and use. It does not matter whether a microphone measures pressure (like the ear) or pressure gradient which is 90 degrees out of phase with it. And the output of the two types can usually be mixed together without problems.

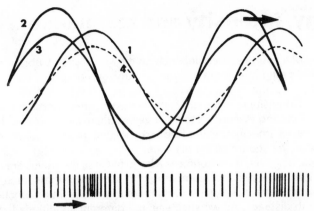

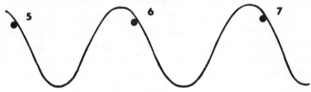

WAVEFORM RELATIONSHIPS 1, Pressure wave. 2, Displacement of air particles. 3, Pressure gradient. 4, Air particle velocity. Note that: (a) pressure is proportional to particle velocity, (b) pressure gradient is proportional to particle displacement, (c) the pressure gradient 'wave' follows a quarter of a wavelength behind the pressure wave. Pressure gradient is said to be 90° *out of phase* with the pressure wave; to lag 90° behind it.

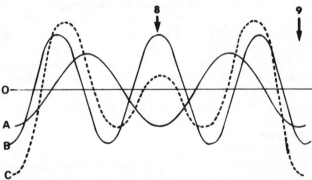

PHASE This is the stage that a wave has reached in its cycle. The points 5, 6, 7 are said to be *in phase* with each other.

ADDING SOUND PRESSURES At any point and at any time the sound pressure is the algebraic sum of all pressures due to all waves passing through that point, i.e. everything above normal pressure is added; everything below it is subtracted. It does not matter what direction the wave is travelling in. If the simple waves A and B are summed the result is shown in the complex waveform C. At 8 there is partial cancellation as pressure in wave A coincides with rarefaction in curve B. At 9 the rarefactions coincide; they reinforce each other.

Energy, intensity and resonance

The energy of a sound source depends on the amplitude of vibration: the broader the swing the more power (energy output per second) it is capable of producing.

The sound intensity at any point is the acoustic energy passing through unit area per second. According to the inverse square law which applies to *spherical waves,* the intensity of sound radiated from a point source diminishes as the square of the distance.

In practice, however, if the source is a large flat area the waves are not at first spherical: for a distance comparable to the size of the source they, too, are flat or *plane waves,* and in this region there is little reduction in intensity with distance. As we shall see, the different conditions that are met in plane and spherical waves affect the operation of pressure gradient microphones.

How a sound source works

To convert the stored energy of the vibrating sound source to acoustic energy in the air, the two must be efficiently coupled. Objects that are small or slender compared with the wavelength associated with their frequency of vibration do not radiate sound at all well. For example, the prong of a tuning fork and the string of a violin slice through the air without giving up much of their energy: the air simply slips round the sides. So in these cases the sound source is coupled to a wooden panel (or box) with a larger surface.

A freely suspended panel has natural frequencies at which it 'rings' if you tap it. If it does this at the frequency of the tuning fork or string, the panel quickly absorbs energy from the source and so can transmit it to the air. The panel is said to *resonate.*

The violin has panels which must respond and resonate at many frequencies. To achieve this it is made very irregular in shape. Such a panel does not ring or resonate to any particular note: if you tap it it makes a dull, unmusical sound. But if the strings are connected to it (through the bridge of the instrument) the panel oscillates in forced vibrations. Energy is transferred from string to panel, and from this to the air.

Cavity resonance

Another sort of resonance is found in the *cavity* or *Helmholtz resonator.* If the panel becomes part of a box with a relatively small opening, the air inside resonates at one special frequency (as when you blow across the mouth of a bottle). This is useful for the sound box of a tuning fork, but in a violin it causes a 'wolf tone' which the violinist treats with particular care.

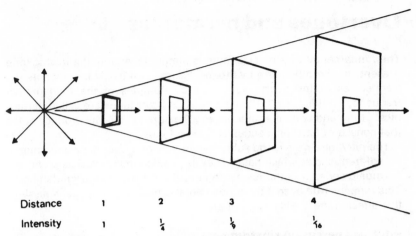

Distance	1	2	3	4
Intensity	1	$\frac{1}{4}$	$\frac{1}{9}$	$\frac{1}{16}$

SOUND INTENSITY is the energy passing through unit area per second. For a spherical wave (i.e. a wave from a point source) the intensity dies off very rapidly at first. The power of the source is the total energy radiated in all directions.

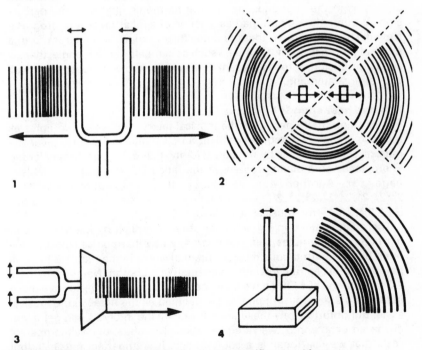

1

2

3

4

TUNING FORK   1 and 2, Each fork vibrates at a specific natural frequency, but held in free air radiates little sound. 3, Placed against a wooden panel, the vibrations of the tuning fork are coupled to the air more efficiently and the fork is heard clearly. 4, When the tuning fork is placed on a box having a cavity with a natural resonance of the same frequency, the sound radiates powerfully.

17

When a single note is played on a musical instrument
many frequencies are produced at the same time.

Overtones and harmonics

The fundamental of a musical note is almost invariably the lowest tone present; all the others are overtones. For many musical instruments the frequencies of the overtones are exact multiples of the fundamental frequency. In this case the fundamental and overtones are called harmonics. String and wind instruments and the human voice all produce notes that consist of harmonic series of related tones.

The *pitch* of a note is the subjective quality of a frequency or combination of frequencies which determines its position in the musical scale. In a harmonic series it is the lowest tone of the series, i.e. the fundamental. This remains true even if the fundamental is weak or completely missing: this affects tone quality, but not pitch.

Pitch and percussion instruments
Many percussion instruments produce notes of indefinite pitch: most of the overtones are not exact multiples of the fundamental. In some, such as tympani, the fundamental is powerful enough to suggest a definite pitch even though the overtones are without harmonic quality. Many percussion instruments are arranged as a series of similar objects of progressively decreasing size, so that the notes they produce when struck form a musical progression; but in others such as cymbals or the triangle, there is such a profusion of tones present that the sound blends reasonably well with almost anything.

How we hear sounds as music
The ear (and brain) perceives the musical interval between two notes as the ratio of their two frequencies. The simplest ratio is 1:2, an octave. The musical scale proceeds by frequency jumps that get progressively larger. The octaves on a piano progress in the ratios 1:2:4:8:16:32:64:128 (see page 13). Another way of writing this is in powers of two, $2^0:2^1:2^2:2^3:2^4:2^5:2^6:2^7$, which leads to a simpler progression: 0,1,2,3,4,5,6,7 – the musical scale is *logarithmic*.

The brain is aided in its judgment of relative pitch by the presence of harmonics in the notes being compared; where harmonics coincide, the brain hears a harmonious musical interval. Notes in the ratio 1:2 have all harmonics (except the lower fundamental) in common. Notes in the ratio 2:3 (an interval of a fifth) have many harmonics in common, as have those in the ratio 3:4 (a fourth); and so on, defining all the notes in the musical scale. But unfortunately these are not exactly the intervals you get if you divide an octave into twelve equal ratios. The equal or 'well-tempered' scale that we use today is a compromise: it is imperfect musically but makes it easier to change key.

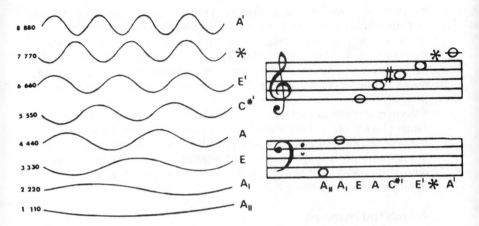

THE FIRST EIGHT HARMONICS OF THE MUSICAL NOTE A_{11} The
fundamental is the first harmonic, the first overtone is the second harmonic
and so on. They are all notes in the key of the fundamental except for the
seventh harmonic which is not a recognised musical note at all: it lies
between G and G♭.

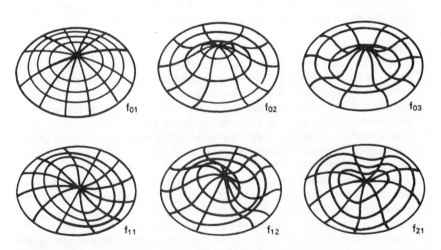

VIBRATION OF A DRUMSKIN   The suffixes indicate the number of radial and
circular lines (nodes) where there is no motion. The first circular node is at
the edge where the membrane is clamped. The overtones are not
harmonically related. If $f_{01} = 100$Hz the other modes of vibration shown here
are: $f_{02} = 230$; $f_{03} = 360$; $f_{11} = 159$; $f_{12} = 292$; $f_{21} = 214$Hz.

Musical acoustics: strings

If a string is bowed, struck or plucked in the middle, the fundamental and odd harmonics, which all have maximum amplitude in the middle of the string, are emphasised; while the even harmonics, which have a node in the middle of the string, are lacking. Exciting the string anywhere near the centre produces strong lower harmonics; exciting it near the end produces strong upper harmonics. The tone quality of the harp is varied in this way.

A violin string is bowed near – but not too near – one end, so that a good range of both odd and even harmonics is excited. But some will be weaker than others. For example, if a string is bowed or struck at approximately one-seventh of its length it does not produce the seventh harmonic, an advantage because this is an overtone that is 'out of tune' with those of other notes that are being played.

Attack and envelope

The method of excitation helps to give an instrument its particular quality. The way that a note starts is called the *attack,* and a microphone must be capable of responding to the rapid and erratic transients that may be present in this part of a note. Early keyboard instruments where the string is plucked are very different in character from the piano, where the string is struck by a felted hammer.

Another quality which defines the instrument is the way in which the note changes in volume as it progresses, i.e. its *envelope.* Here the violin and its family, in which the strings are excited continuously, are characteristically different from the piano, harp and harpsichord.

Tone quality

The tone quality of the individual instrument within a family – e.g. violin, viola, 'cello or bass – is defined by the qualities of the resonator; and most particularly by its size. A large resonator responds to and radiates lower notes than a small one, while the sound box of a violin is too small to radiate the fundamental of its lowest note (the low G) effectively. There is also a marked difference in quality between the lowest notes (very rich in harmonics), and the highest (strong in lower and middle harmonics, but relatively thin in tone colour).

As we have seen, the point at which the string is excited emphasises some frequencies at the expense of others; and the shape and size of the resonator modifies this still further. This characteristic of a resonator is called its *formant.* Formants are obviously a virtue in music, but in audio equipment they are called 'an uneven frequency response' and are usually regarded as a vice.

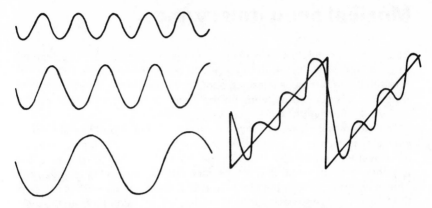

SAWTOOTH WAVEFORM   The first three harmonics add together to produce a first approximation to a 'sawtooth', the waveform that is produced by drawing a bow across the string of a violin. The string is dragged and then slips, repeatedly.

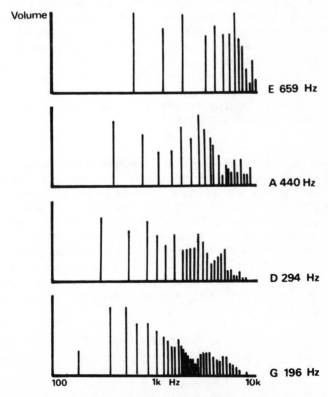

Volume

E 659 Hz

A 440 Hz

D 294 Hz

G 196 Hz

100   1k   Hz   10k

THE OPEN STRINGS OF THE VIOLIN: the relative intensity of the harmonics.

21

Musical acoustics: wind

The air column inside a wind instrument may be excited by the vibrations in turbulent air at the edge of an orifice (as in the flute), by one or two reeds in the mouthpiece (clarinet, oboe) or by the player's lips (brass). The note that is produced depends on the dimensions of the resonator itself, and on the formation of standing waves within it.

The progressive waves of radiating sound are converted into stationary waves by reflection at a boundary. At a solid, rigid surface there can be no air movement perpendicular to it: the incident and reflected waves add together in such a way that air displacement at the wall is always zero. This is called a node in the standing wave: there will be further nodes at the first, second, and each successive complete wavelength from the wall; while at the intervening half wavegths there will be antinodes, places where there is maximum air displacement.

Behaviour of standing waves
In practice, not all of the wave is reflected: the surface absorbs some of the sound, and transmits some through to the other side, but in the air column of a wind instrument the standing wave dominates the pattern. In fact, one or both ends of the air column of a musical instrument may be open, but reflections still occur because the pipe is narrow compared with the wavelength of most of the sound in the column: the air pipe forms a piston which is simply not big enough to drive the outside air efficiently. In this case the energy has to go somewhere: it is reflected repeatedly back along the column, building up a powerful standing wave, whose pitch is defined by the length of the column.

Harmonics in an air column
Some instruments have air columns that are open at one end; some are effectively open at both. A column that is open (or closed) at both ends has a fundamental wavelength which is twice the length of the pipe, and produces a full harmonic series. An air column that is open at one end and closed at the other behaves differently: the fundamental is four times its length, and only the odd harmonics are produced.

In orchestral instruments the length of the column may be varied continuously (as in the slide trombone), by adding discrete extra lengths (trumpet and French horn) or by opening and closing holes in the body of the instrument (woodwind). The formant of an instrument depends on the shape of its body and bell (the open end).

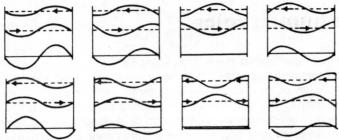

STATIONARY WAVE formed as two progressive waves move in opposite directions, successively reinforcing and cancelling each other. The combined wave has twice the amplitude of the original waves.

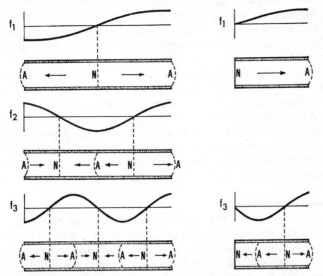

VIBRATION OF AIR COLUMN   *Left:* for a pipe open at both ends the fundamental frequency f_1 is approximately twice its length. The overtones are multiples of the fundamental: $f_2 = 2f_1$; $f_3 = 3f_1$ and so on.
Right: for a pipe that is closed at one end the fundamental frequency f_1 is four times its length. The even harmonics are absent: the first overtone is $3f_1$ the next $5f_1$ and so on.

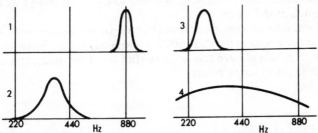

INSTRUMENTAL FORMANTS   1. Oboe. 2. Horn. 3. Trombone. 4. Trumpet. The formants, imposed by the structural dimensions of parts of the instruments provide an essential and recognizable component of their musical character.

23

The human voice

The special qualities of the human voice lie – in part – in the precise control of pitch that is given by the vocal cords, combined with flexibility of the cavities of the mouth, nose, and throat which are used to impose variable formant characteristics on the sounds already produced.

These formant characteristics, based on cavity resonance, are responsible for vowel sounds and so make an important contribution to the intelligibility of speech. Most of this variability is expressed between 200 and 2700Hz, the range of telephone transmissions.

Other characteristics of speech include sibilants and stops of various kinds which, together with the formant resonances, provide all that is needed for high intelligibility. These extend the range to about 8000Hz. A whisper, in which the vocal cords are not used, may be perfectly clear and understandable.

The vibrations produced by the vocal cords add volume, further character, and the ability to produce song. For normal speech the fundamental varies over a range of about twelve tones and is centred somewhere near 145Hz for a man and 230Hz for a woman. As the formant regions differ little, the female voice therefore has less harmonics in the regions of stronger resonance; so its quality may be thinner (or purer) than a man's. For song the fundamental range of most voices is about two octaves – though, exceptionally, it can be much greater.

Microphones for speech

For a microphone and other sound equipment that can cope with orchestral music, the human voice produces no problems of frequency range, nor of volume except for the plosive 'p' that can blow a diaphragm beyond its normal working range when some people talk directly into it.

As we shall see, the demands of intelligibility require that there must be less indirect (reverberant) sound than for musical instruments; and this in turn may create further problems due to working close to a directional microphone. But apart from this, the main question with microphones for speech is whether a very wide frequency range is desirable: this adds not only to the cost of the microphone but also to the noise picked up at the extreme ends of the range.

Sometimes intelligibility is increased by using a microphone with a peak (a stronger response) in the 6000-8000Hz frequency range. Some older or cheaper microphones have this anyway. But this may also enhance the natural sibilance of some voices.

VOCAL CAVITIES   1, The lungs. 2, The nose. 3, The mouth (this is the most readily flexible cavity and is used to form vowel sounds). 4, The pharynx (above the vocal cords). 5, Sinuses. These cavities produce the formants which are characteristic of the human voice, emphasising certain frequency bands at the expense of others.

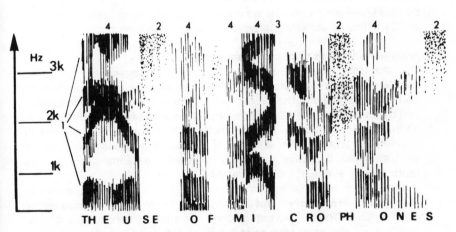

HUMAN SPEECH analysed to show formant ranges. 1, Resonance bands. 2, Unvoiced speech. 3, Vocal stop before hard 'a'. There would be similar break before a plosive 'p'. 4, Voiced speech. These formants are unrelated to the fundamentals and harmonics, which would be shown only by an analysis into much finer frequency ranges than have been used here. This example is adapted from a voiceprint of the author's voice. Each individual has distinct characteristics.

Sound volume and the ear

Every time you *double* the intensity (or the energy) of a sound your ear hears the *same* increase in loudness. Intensity is a physical characteristic of sound in air. As intensity goes up in progressively increasing jumps, 1:2:4:8:16:32 and so on, the ear hears this as equal increments of loudness, 1:2:3:4:5:6. Because of this special characteristic of the human ear, it is not very convenient to think in terms of intensity or energy, so instead a new measure of sound volume is defined. This is the *decibel* (dB).

In mathematical terms: the volume in decibels is ten times the logarithm of the ratio of intensities. For non-mathematicians it is enough to know that a regular increase in volume 1:2:3:4:5:6 described in decibels will sound like a regular and even increase in volume. And a decibel is a unit of a convenient size: it is about as small an increase in volume as we can hear in the most favourable circumstances.

Intensity and volume
As it happens, the ratio of intensities in 3dB is about 2:1. This is convenient to remember, because if we double up a sound source we double the intensity (at a given distance). So if we have one soprano bawling her head off and another joins her singing equally loudly the sound level will go up 3dB (not all that much more than the minimum detectable by the human ear). But to raise the level by another 3dB two more sopranos are needed. Four more are needed for the next 3dB – and so on. Before we get very far we are having to add sopranos at a rate of 64 or 128 a time to get any appreciable change of volume.

Human hearing at different frequencies
The ear does not measure the volume of all sounds by the same standards: the ear is more sensitive to changes of volume in the middle and upper frequencies than in the bass. So loudness (a subjective quality) and the objectively measurable volume of a sound cannot be the same at all frequencies. Loudness in *phons* is taken to be the same as volume in decibels at 1000Hz. The expected loudness of a sound can then be calculated from the actual sound volume by using a standard set of curves representing 'average' human hearing.

The lower limit is called the threshold of hearing. It is convenient to regard the average lower limit of human hearing at 1000Hz as zero on the decibel scale. (There is no natural zero: the arbitrary zero chosen corresponds to an acoustic pressure of 2×10^{-5} newtons per square metre.)

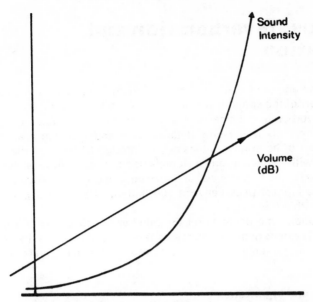

INTENSITY AND VOLUME   As sound intensity rises exponentially the
corresponding volume in decibels (as heard by the ear) rises linearly.

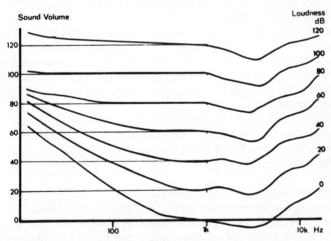

EQUAL LOUDNESS CONTOURS   (Fletcher-Munson curves). The scale on
the left is volume; that on the right is loudness; the two are equal at 1000Hz.
These typical contours are usually described as 'for normal ears', but hearing
that is regarded as normal may differ substantially from this (though the
threshold of hearing should be reasonably close to the lowest curve). Note
that the ear can less easily distinguish differences in volume at low
frequencies – except at very high volume, where an increase in physical
sensation makes it easier. The upper limit of hearing is called the threshold of
pain and is usually regarded as being somewhere between 100–120dB. (But
in fact a low-frequency noise field of 135dB is not necessarily painful, though
it is uncomfortable: it is difficult to breathe.)

Studios: reverberation and coloration

Sounds in an enclosed space are reflected many times, with some (great or small) part of the sound being absorbed at each reflection. The rate of decay of reverberation defines a characteristic for each studio: its reverberation time. This is the time it takes for a sound to die away to a millionth part of its original intensity (i.e. through 60dB). Reverberation time varies with frequency, and a studio's performance may be shown on a graph for all audio frequencies. Or it may be simply given in round terms for, say, the biggest peak between 500 and 2000Hz; or at a particular frequency within that range.

Reverberation time depends on the number of reflections in a given time, so large rooms generally have longer reverberation times than small ones. This is not only expected but also, fortunately, preferred by listeners.

Coloration and eigentones

In some studios coloration may be heard. This is the selective emphasis of certain frequencies or bands of frequencies in the reverberation, caused by successive absorption of other frequencies at each reflection. This is most noticeable in smaller rooms.

Parallel walls also give rise to eigentones, resonance at frequencies corresponding to the studio dimensions. If the main dimensions are in simple ratios to each other these may be reinforced. Adequate diffusion is an important quality of a good studio. The more the wave front is broken up, the more smooth the decay of the sound becomes, both in time and in frequency content.

A hand clap can be used as a rough guide to both reverberation time and studio quality: for example in a speech studio, the sound should die away quickly but not so fast that the studio sounds muffled or dead. And there must certainly be no 'ring' fluttering along behind it.

Reverberation times for music studios

For a music studio, listening tests show that the preferred reverberation time depends on size: the larger the studio, the longer the time, but the number of sound reflections remains about the same. For monophonic radio and recording work the preferred ideal reverberation time seems to be about a tenth of a second less than for music heard 'live'. Some authorities suggest a slight rise in the bass, but for its studios the BBC aims for a response that is flat at all frequencies.

BBC listening rooms are acoustically engineered for a response of 0.4 second up to 250Hz, falling gradually to 0.3 second at 8000Hz.

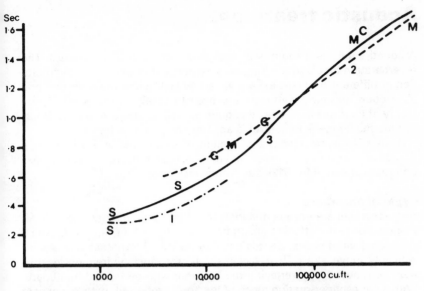

REVERBERATION TIMES FOR SOUND STUDIOS   1 and 2, The generally accepted optimum reverberation times for speech (S) and music (M) (suggested by Beranek). 3, BBC experience of good studios is shown in the central line. This includes: G. General-purpose studios which may have acoustics which are relatively live at one end and dead at the other. C. Concert halls.

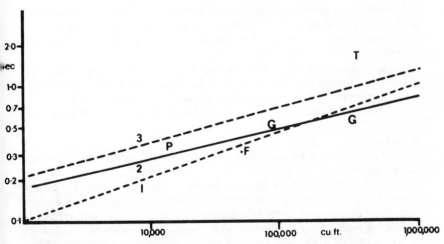

REVERBERATION TIMES FOR TELEVISION STUDIOS. 1, Lowest practicable limit. 3, Highest acceptable figure. 2, Normal design aim for BBC studios: P. Typical small presentation studio. G. General-purpose studio. Difficulties encountered by using existing buildings are shown by F, an old film studio, which was found to be uncomfortably dead, and T, a converted theatre, which was much too live.

29

Acoustic treatment

A sound studio is a room with designed – or defined – acoustics. The reverberation time and the frequency response (the amount of reverberation at different frequencies) are varied by controlling the reflection and absorption of sound. Diffusers are used to break up the wave fronts; plenty of hard furniture does this quite as well as irregular wall surfaces. But the main means of control is absorption in various forms.

Absorption coefficient is the fraction of sound that is absorbed at a particular frequency. It takes a value between 0 and 1 and unless otherwise stated is for 512Hz at normal incidence.

Types of absorbers
Soft absorbers are porous materials in which sound energy is lost by air friction. For really efficient absorption it has to be four feet thick, which takes up a lot of space. So it is usual to lay it in thicknesses that absorb well only above 500Hz. Excessive absorption at very high frequencies may then be reduced by overlaying the absorber with perforated hardboard. With 5 per cent perforation much of the 'top' is reflected; with 25 per cent perforation a high proportion is absorbed.

Damped cavity absorbers use the principle of the Helmholtz resonator, in which the mass of air in the mouth of a bottle vibrates against the 'spring' of the air inside. Damped cavity resonators are sometimes used to cut down sharp peaks in studio reverberation – such as may be caused by dimensional resonances.

Membrane absorbers are often used to reduce low frequencies: a panel with a broad middle- to low-frequency response is driven like a piston by the sound pressure, and as the movement is damped the energy is lost. In practice they do not always work well.

An all-purpose combination absorber has been used by the BBC with success. It consists of standard two-foot square units that are easy to construct and simple to install without supervision (see opposite).

The best – and cheapest – sound absorber on the floor is a thick carpet.

Screens
Screens are free-standing diffusers or diffuser-absorbers, typically about three feet wide by six feet high. They are not very efficient absorbers, as the padding is usually too thin. In the dead pop-music studio, Perspex screens provide local high-frequency reinforcement. This helps musicians hear their own sound, helps to separate instruments for easier control of levels, improves diffusion and may also reduce low-frequency resonance in the studio as a whole.

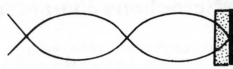

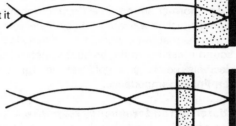

SOFT ABSORBERS Padding applied close to a reflecting surface is efficient only for sufficiently short wavelengths. The thicker it is the wider the range of frequencies that it will absorb. Thin padding (or, for example, heavy drapes) is more effective over a wide range of frequencies if it is away from the reflecting surface.

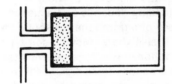

DAMPED CAVITY ABSORBER This employs a Helmholtz resonator responding to a particular frequency, which is then absorbed.

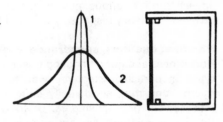

MEMBRANE ABSORBER A panel over a box. Unlike the cavity resonator (which has a single sharply tuned resonance), 1, the panel which is fixed at the edges has a very broad response, 2. If it is damped by weighting it with a material such as roofing felt, it will absorb sound.

COMBINATION ABSORBER, size: 2ft x 2ft x 7in. The back and sides of the box are of plywood. The interior is partitioned by hardboard into cavities of 1ft x 1ft x 6in. On the front is 1in of heavy density rockwool, covered by perforated hardboard. With 5 per cent perforation the peak of absorption is at 90Hz. With 20 per cent perforation there is wide band absorption with less at low frequencies.

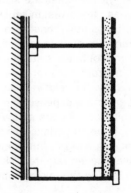

31

Microphone characteristics

A microphone is a device for turning the acoustic energy of sound into electrical energy. The main types in use are condenser, moving-coil and ribbon for high-quality work, and crystal and carbon for other uses. Microphones described as 'dynamic' are generally of the moving-coil type; a 'velocity' microphone is of the ribbon type.

For normal sound levels a microphone should produce an electrical signal which is well above its own electrical noise level, i.e. it should have a good signal-to-noise ratio. Sensitivity is the voltage produced per unit of sound pressure. In the present system of units this is generally quoted in decibels below one volt per newton per square metre for a given terminating impedance.

Distortion and frequency response
For normal sound levels the signal must be substantially undistorted, i.e. the electrical waveform must faithfully follow the sound waveform, and it must do this over a wide range of volumes. Distortion caused by overloading depends on the use to which the microphone is put: close to the bell of a trombone the sound level may be 60dB higher than that met in use elsewhere. Moving-coil types withstand overload very well. The microphone taken together with its associated equipment should respond almost equally to all significant frequencies present. What is 'significant' depends on the sound source: different microphones may be used for speech and for various musical instruments. Sometimes deviations from an even frequency response are desirable, and can be used constructively.

Handling qualities, appearance, cost
Robustness and good handling qualities are other essentials. Moving-coil types in particular stand up well to the relatively rough treatment a microphone may accidentally suffer outside a studio, to handling by interviewers or pop singers, or to rustle when attached to clothing. Condenser microphones are also used for all of these purposes. Ribbon microphones rate lowest in sensitivity to wind noise. In windy conditions all microphones may benefit from the use of a suitable windshield.

Appearance matters in television, while size and weight are important when the microphone is attached to clothing.

Crystal microphones are very cheap, and good moving-coil microphones may be relatively so. Condenser microphones (other than electrets) are more expensive; ribbons vary: some are expensive.

Some microphones are sold with a choice of impedances. Nominally, the microphone impedance should be matched with the input to the equipment following it, but sometimes a mismatch – low into high – is permissible or even required. Non-technical users should take advice on this when buying the microphone.

CRYSTAL MICROPHONE 1, Light rigid diaphragm. 2, Crystal bimorph: two slices of a crystalline material cut so that their axes are crossed. As the bimorph is twisted a voltage (called a piezo-electric voltage) is generated between foil plates on the surfaces of the sandwich. Not used for high-quality work. Now largely displaced by low-cost electrets.

RIBBON MICROPHONE
1, Corrugated light alloy foil diaphragm. 2, Permanent magnet with pole-pieces above. 3, Transformer. The output voltage is generated as the ribbon diaphragm vibrates in the magnetic field.

MOVING COIL MICROPHONE
1, Diaphragm. 2, Coil fixed to diaphragm. 3, Permanent magnet. Movement of the coil over the fixed centre-pole produces the output voltage. A robust microphone.

CARBON MICROPHONE
1, Diaphragm. 2, Granular carbon button. 3, Battery. 4, Transformer. Pressure on the diaphragm changes the resistance of the carbon, to which a polarising voltage has been applied. Not used for high-quality work.

CONDENSER (ELECTROSTATIC) MICROPHONE 1, Light flexible membrane diaphragm. 2, Rigid backplate. 3, Cable to head amplifier and polarising voltage supply. Air pressure varies the capacitance between diaphragm and backplate. This is converted to an output voltage in the amplifier. Expensive, but can give very high sound quality.

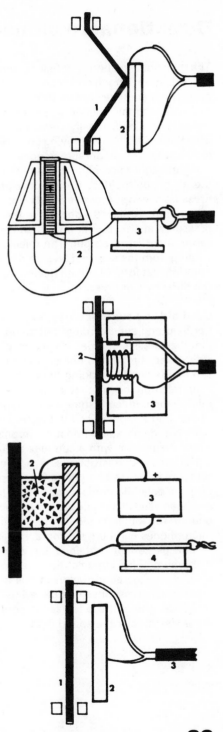

Directional response

The directional response or field pattern of a microphone is illustrated in a polar diagram, in which sensitivity is plotted against the angle from the axis. The main types are as follows:

Omnidirectional microphones measure the pressure of the air in the sound wave and, ideally, respond equally to sound coming from all directions. The simpler moving coil, crystal and carbon microphones work in this way. The diaphragm is open to the air on one side only.

Figure-eight microphones are bi-directional. The diaphragm is open to the air on both sides: it measures pressure gradient (the difference in pressure at two successive points along the path of the sound wave). If the microphone is placed sideways to the path of the sound, it compares the pressures at two points on the same wave front and, as these are the same, there is no output. The microphone is therefore 'dead' to sound coming from the side and 'live' to that approaching one face or the other. The simplest form of ribbon microphone works in this way, responding to the difference in pressure on the two faces of a strip of foil.

Cardioid and switchable types

Cardioid and supercardioid microphones are unidirectional in the sense that they are live on one side and relatively dead on the other, but the live side is very broad and undiscriminating. A heart-shaped polar response can be obtained by adding the output of a pressure (e.g. simple moving coil) microphone to that of a pressure gradient (e.g. simple ribbon) microphone. Indeed, some earlier types of cardioid did just that, with a ribbon and a moving coil in the single case. Today, many cardioids still have two diaphragms (most commonly in condenser microphones) or a single diaphragm with a complex acoustic network behind.

The cardioid microphone is a special case of the combination of pressure and pressure gradient principles. By varying the proportions, a range of polar diagrams is possible. Where a combination of microphones has been used in a single case, it has been possible to add a switch so that one or the other could be switched off. Twin-diaphragm condenser microphones can be even more versatile, as a whole range of different responses can be obtained by varying the polarising current of the diaphragm. The supercardioid response discriminates most strongly against reverberation; between this and the pure cardioid lies the most unidirectional response (see page 47). For an even more highly directional response, one of these is used in combination with a device to focus or otherwise modify the sound field.

POLAR DIAGRAM: PERFECT OMNIDIRECTIONAL RESPONSE The scale from centre outwards is sensitivity, measured as a proportion of the maximum response (which is taken as unity). An alternative scale is in decibels. A scale with 0 at the outside and −25 to −35 near the centre gives a diagram that is similar over the main working range.

PERFECT FIGURE-EIGHT RESPONSE (BI-DIRECTIONAL) The response is similar in the plane through the 0°–180° axis; that in the plane at right angles to the 90°–270° axis is always zero. At 60° off axis the output is reduced to half. The response at the back of the microphone is opposite in phase.

PERFECT CARDIOID RESPONSE is the sum of omnidirectional and figure-eight pick-up when the maximum sensitivity of the two is equal. The front lobe of the 'eight' is in phase with the omnidirectional response and so adds to it; the back lobe is out of phase and is subtracted.

TYPICAL SUPERCARDIOID RESPONSE This is a more directional version of the cardioid microphone. It is one of a continuous range of patterns that can be obtained by combining omni- and bi-directional response in various proportions.

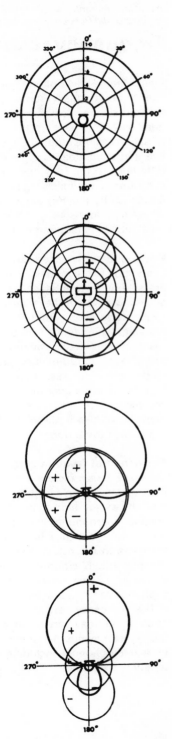

35

In real life, microphones can match up only approximately to the ideal flat frequency response.

Frequency response in practice

Even when a microphone has a response that is almost flat on its main axis there is usually some degradation at the sides or back: the polar diagram changes with frequency. These changes are most marked on the dead sides of directional microphones, but they also matter least in these directions as the output is so low. But in the intermediate ranges between live and dead sides a slight change of direction can make a substantial difference to the frequency response to direct sound, as a study of the diagrams on the following pages will show. This can often be used to help a balance by angling the microphone to reduce high-frequency content where it is obtrusive, or pointing it directly at the source when increased intelligibility is required in a noisy background.

Factors affecting frequency response

In addition to the mode of operation (pressure or pressure gradient), there are several factors which affect frequency response. These include: the shielding effect of the microphone body, which reduces high-frequency response to sound from that direction; reflection within the diaphragm housing, which tends to produce an erratic response at high frequency; resonances of the air cavities between casing and diaphragm, which may produce a high-frequency peak; resonances in the cavity behind the diaphragm or in pressure equalisation channels; cancellation of extreme high frequencies over the surface of a diaphragm when wavelengths are comparable with its size; a tendency to form standing waves in front of the diaphragm, also in the highest frequencies; the resonance of the diaphragm itself (on a ribbon this is at a low frequency).

The effect of frequency response

Many of these factors are used constructively in the design of the microphone, often resulting in a slight increase in sensitivity along and near the axis at high frequency (at, say, 4000Hz and above). But older or cheaper microphones with large diaphragms (e.g. of 1½in diameter) show these high-frequency peaks to excess. Typically there may be a spiky response to the octave from 5000–10 000Hz, perhaps with a peak of as much as 7dB. Microphones with small diaphragms (typically less than ½in diameter) have a more even response. Among these, condenser microphones can be particularly good, as the diaphragm is so light.

The frequency response of one sample of a particular high-quality microphone may vary by several dB from another, but (except for stereo pairs) this may not matter very much. In the complexity of sound received by a microphone – including reverberation – such small variations are barely perceptible.

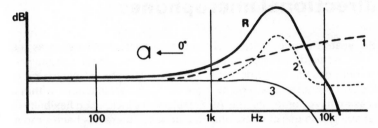

HIGH-FREQUENCY EFFECTS on a microphone diaphragm. The response R at 0° (i.e. on axis) is the sum of several components including: 1, Obstacle effect (which sets up standing wave in front of diaphragm). 2, Resonance in cavity in front of diaphragm. 3, Inertia of diaphragm as its mass becomes appreciable compared with high-frequency driving force.

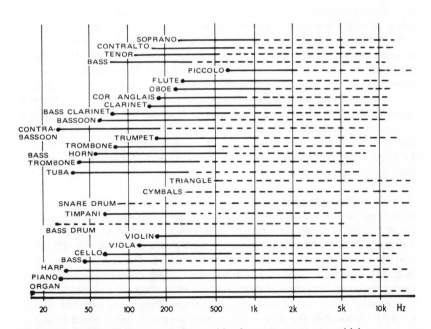

FREQUENCY RANGES   A microphone with a frequency response which matches that of the sound source will not pick up noise and spill from other instruments unnecessarily. The lowest frequency for most instruments is clearly defined, but the highest significant overtones are not. At high frequencies, an even response is usually considered more valuable than a matched one.

Figure-eight microphones can be used to control acoustics as well as to favour particular sound sources.

Bi-directional microphones

Over the years the popularity of bi-directional microphones such as the single ribbon has diminished as the cost of both condenser and moving coil microphones has come down; as the condensers have become more versatile and more reliable; and as more complex combinations within a single casing have come into use. As a result of this increasing flexibility in the availability and use of directional microphones there has been a move away from the operational procedures that employed bi-directional microphones, and a tendency towards using one microphone for each source. In addition the quality of response of the ribbon microphone itself has been overtaken by the improvements in cardioids using other principles: the response of the ribbon is neither so smooth nor so extensive. Nevertheless it remains valuable for a limited but important range of uses such as the human voice and certain musical instruments (notably strings) whose frequency characteristics happen to fit in with those of the ribbon microphone.

Control of acoustics
In bi-directional use the working area is wide – including about 100° on either side of the microphone – which allows room for a number of people (four sitting or six standing) at about two feet from the microphone. The directional response allows the use of studios or rooms with fairly lively acoustics without these becoming too dominant. To an omnidirectional microphone, reverberation comes from all sides; to the bi-directional microphone it comes only from the double cone at front and rear. For the same acoustic the speakers must be much closer to an omnidirectional microphone to achieve a similar balance of direct and indirect sound; slight movements then have a much greater effect on the ratio. The bi-directional microphone therefore allows more flexibility and greater subtlety in the use of acoustics.

Avoiding phase problems
When two bi-directional microphones are used near each other they must be *in phase*. If they are not, there is cancellation of direct sound for sources which are at similar distances from the two. To avoid this there is usually some visual indication of which side is the 'front'. When two bi-directional microphones are used for separate sources they can often be arranged so that each is dead-side-on to the other's source.

The polar response of a ribbon microphone that is set with the ribbon upright approaches the ideal in the horizontal plane, but is more erratic in the vertical plane, due to phase cancellation. As with most such variations in response this, if known, can be used constructively.

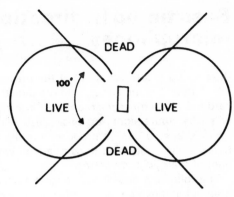

FIGURE-EIGHT MICROPHONE
Working areas. An arc of approximately 100° on each side is 'live'.

RIBBON MICROPHONE
Much of the casing is taken up by the magnet system, which in early ribbon microphones had to be substantial in order to provide reasonable sensitivity.

Frequency response of ribbon microphone. The response at 60° to the axis is 6dB down as predicted by theory, but begins to depart from the common curve at about 2000Hz, where high-frequency effects begin. Those for variation in the vertical plane (v) are more pronounced than those for variation in the horizontal plane (h).

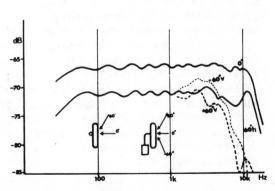

39

Bass tip-up is an increase in bass response that occurs when a sound source is within about two feet from a directional microphone.

Bass tip-up in directional microphones

Bass tip-up is a direct result of the pressure gradient mode of working, which in turn is fundamental to the directional response of bi-directional, cardioid and similar microphones. It is found in its most extreme form with the figure-eight response, less with cottage-loaf, less still with cardioid and so on – in proportion to the diminishing importance of the figure-eight component in comparison with the omnidirectional component of their polar diagrams.

The following description is for a ribbon microphone, pure pressure-gradient and exhibiting the effect in its strongest form.

Tip-up occurs when the pressure gradient microphone is in a field of spherical waves and if it is close enough to the source for there to be a substantial reduction in intensity within the distance that the wave has to travel to get from front to back of the diaphragm. This loss of intensity creates a pressure difference which is added to the normal pressure gradient that is due to the phase difference. The effect is only appreciable at low frequencies where the phase difference is small compared with the fall in sound intensity.

Uses of bass correction
The response of some directional microphones falls off below 200Hz; in these, bass tip-up may be used to restore a level response. In cases where the response is more nearly level, 'bass-cut' (actually bass-correction) is required. This may be applied either in the control channel following the microphone or sometimes by switching in a filter built into the microphone itself. One result of bass-cut is a reduction in studio reverberation at low frequencies, which may be used to compensate for small-room resonances or 'boominess'.

Working distance
Without bass correction the minimum working distance may be as much as 2ft or as little as 1ft or less, depending on the microphone. With correction the optimum working distance is reduced, but any movements backward and forward of more than a few inches then produce a noticeable change in bass response even if the actual volume is controlled to compensate for the change in intensity. There is also (in this case) a variation in the ratio of direct to indirect sound.

Bass tip-up is more troublesome on some voices than others. Women, with a higher fundamental frequency, can often speak at less than the recommended working distance without distortion becoming apparent. For whispers it is often possible to speak very much closer, because the main frequency content of a whisper is generally much higher than that of voiced speech.

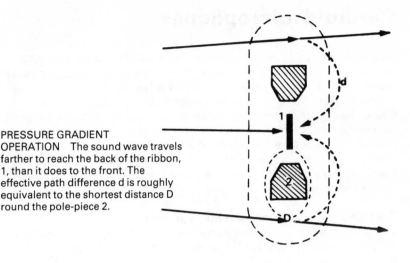

PRESSURE GRADIENT OPERATION The sound wave travels farther to reach the back of the ribbon, 1, than it does to the front. The effective path difference d is roughly equivalent to the shortest distance D round the pole-piece 2.

BASS TIP-UP An extreme case of increase in bass with the sound source at various distances from a directional microphone. This is for a bi-directional microphone (the ribbon shown on p.39). Other directional microphones, including cardioid and cottage-loaf, show the same effect but in lesser degree.

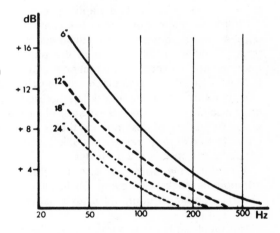

COMPENSATION FOR CLOSE WORKING Some microphones have a reduced bass response which automatically compensates for close working: with the microphone at a certain distance from the source, its response for direct sound is level.

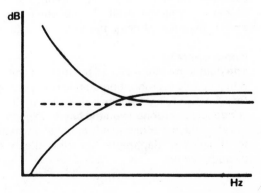

41

Microphones that are live on one side only.

Cardioid microphones

With true cardioid microphones the working area lies within a broad cone including about 160° on the live side. At the sides the response diminishes rapidly, and at the back, where the output is low, the frequency response is erratic. Bass tip-up occurs with close working, though less than for figure-eight microphones. Some cardioids tend to degrade to omni-directional response at low frequencies, and this considerably reduces the tendency to bass tip-up. However, the degradation may itself depend on distance as well as frequency; so it is necessary to study the manufacturer's details first, and then learn the possibilities and eccentricities of each model from actual use.

Combination and phase shift microphones
The original way of producing a cardioid, by combining a moving coil and a ribbon in a single case, has fallen into disuse, but most modern cardioids are even more complex. The most commonly used method today employs a condenser with two diaphragms – of which more later.

Directional microphones may also contain a pair of moving coils back to back, or two ribbons one behind the other. By connecting them electrically through networks which change the phase of the signal, a response intermediate between omnidirectional and figure-eight can be achieved. In these cases the second diaphragm is behind the first on the main axis, and therefore has a slightly different response; and the polar diagram must also depend on the relative sizes of the two signals. There is so much scope for variety of design detail that it is impossible to generalise; it is better to study the characteristics of individual types as they are produced for particular specialised purposes.

A cardioid response can also be obtained in a single diaphragm microphone, such as a moving coil, by allowing sound to reach the back of the microphone through an acoustic labyrinth so that it arrives with a change of phase.

The diaphragm or diaphragms may be mounted in the microphone casing in two main ways. One, 'end-fire', has the main axis in line with that of the microphone itself; in the other, more commonly employed in switchable microphones, the axis is at right-angles to it.

Boom operation
The double-moving-coil and the moving-coil phase-shift microphones are both suitable for mounting on television or film booms. They are satisfactory for following speech but for music it is generally better to use a condenser microphone on the boom. Experience will show how vibration-resistant the mounting needs to be; and whether windshields are required to protect the diaphragm from the effects of moving the microphone through the air.

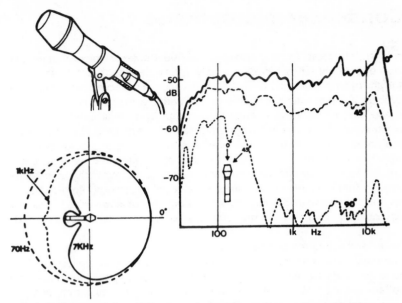

DOUBLE MOVING COIL CARDIOID MICROPHONE This is an example of an 'end-fire' microphone: the 0° axis of the diaphragm is along the axis of the casing.

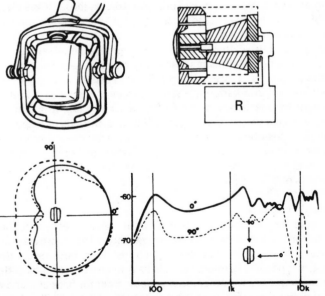

MOVING COIL CARDIOID MICROPHONE suitable for television boom operation. A complex system of acoustic labyrinths is built into the microphone housing: the air reservoir R (shown in the simplified sectional diagram) provides the damping for one of three resonant systems which are used to engineer the response of the microphone.

For very high-quality work condenser microphones have become the most commonly used type.

Condenser microphones

In their simplest form condenser (or electrostatic) microphones are omnidirectional. They have a single thin flexible diaphragm close to the surface of a rigid backplate. There are changes in capacitance between the two, as the diaphragm vibrates in sympathy with the sound pressure. These are converted to changes in voltage in a small amplifier that is either within the same housing as the capsule or nearby. The diaphragm itself may be of very thin metal, perhaps in the form of a gold-sputtered plastics membrane. The backplate has small cavities in it: without these the thin layer of air would be too 'stiff' and would prevent the membrane moving freely. To make the microphone work a polarising voltage has to be applied between diaphragm and backplate, or alternatively may be introduced permanently in manufacture (as in an 'electret'; see page 61).

Double-diaphragm condensers

Some condenser microphones in use today are more complicated. They have two diaphragms, one on either side of the backplate. If the polarising voltage is the same on both diaphragms, the microphone continues to be omnidirectional in operation. With no polarising voltage applied to the second diaphragm, it can be arranged that the microphone has a cardioid response. In this case, the cavities in the backplates are designed as a phase shifting network: as the acoustic pressure travels through the plate to the back of the main diaphragm it changes in phase sufficiently to generate the cardioid response.

Going a stage further, if the second diaphragm is polarised in the opposite direction, a figure-eight response can be obtained. The double diaphragm then acts as a pressure gradient capsule. Finally, if the polarising voltage is made variable a whole range of response patterns can be obtained. In one example, any of nine polar diagrams can be selected at the turn of a switch. In these microphones, the axis of response is at a right angle to the body of the microphone, in order to accommodate the figure-eight mode of operation. Their poorest response is along the line of the body of the microphone: the response from 4000Hz upward is erratic in just this one direction.

Interchangeable capsules

An alternative to the double diaphragm is to have interchangeable capsules fitting on to a common microphone body. These provide for omnidirectional and cardioid responses. They can then both be of the single diaphragm types, the cardioid capsule having an acoustic labyrinth.

Condenser microphones may include a variable filter for bass cut, and a filter to eliminate the radio-frequency signals which are very easily picked up in the stage from the capsule to the head amplifier.

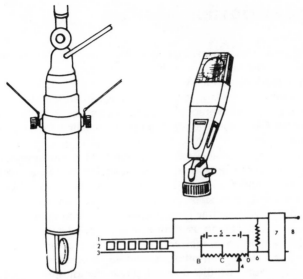

SWITCHABLE CONDENSER MICROPHONE   *Left:* the early version of a high-quality variable response condenser microphone, and *top* a later smaller version that has replaced it. The circuit diagram shows how the response is changed by varying the polarising voltage to one diaphragm. 1, Front diaphragm. 2, Rigid centre plate (perforated). 3, Rear diaphragm. 4, Multi-position switch and potentiometer. 5, Polarising voltage. 6, High resistance. 7, Head amplifier. 8, Microphone output terminals. When the polarisation is switched to position O the voltage on the front and back diaphragm is the same, and above that of the centre plate: the capsule operates as an omnidirectional (pressure) microphone. At B the capsule measures pressure gradient and is bi-directional. At C the polar response is cardioid.

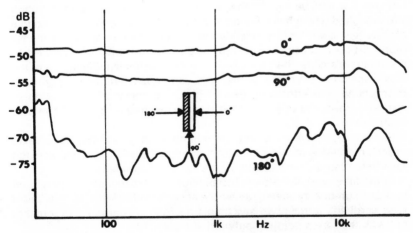

CARDIOID RESPONSE OF HIGH-QUALITY CONDENSER MICROPHONE   The axial response is substantially flat to 15 000Hz and the response at 90° almost so. The irregularities in the 180° curve are unimportant.

Supercardioids

Flexibility of polar response started with being able to choose between omnidirectional and bi-directional types; then between these and the cardioid combination of the two, and so to the switchable condenser microphones with a choice of nine conditions. Given the choice in this way, users rapidly discovered the special value of the range that lay between cardioid and figure-eight: the response called supercardioid or hypercardioid.

We have seen that a directional microphone can favour one particular source or group of sources and discriminate against others, and also that it picks up less studio reverberation so that it can be used further back, giving a more evenly balanced coverage of a widely spread source. The figure-eight has excellent directional properties, but unless the back of the microphone is actually being used for a second sound source, the size of the rear lobe can be an embarrassment: it may pick up an excess of reverberation or sound from other, unwanted sources. In these circumstances, where much of the forward directivity of the bi-directional microphone is required, but *not* its full rear pick-up, the supercardioid response is a distinct improvement. It may be particularly valuable for individual instruments in a multi-microphone music balance, or for single sources in a lively acoustic. For such purposes microphones with a fixed supercardioid response have been specially designed, seeking for each use the best compromise between directivity, discrimination against ambient sound, and bass tip-up.

Bass tip-up turned to advantage
It should be noted that bass tip-up is to be expected with close working, but this too can be turned into a virtue – as with one double-ribbon supercardioid which has a fall-off in the bass from 200Hz. Used close to a source, this lower limit is extended by perhaps half an octave, while for more distant sources (and reverberation) the microphone continues to discriminate against the bass which is always the most difficult part of the sound spectrum to control acoustically. The bass fall-off in this particular microphone also effectively deals with the problem of degradation of frequency response which often affects the low frequencies of directional microphones.

Some inexpensive supercardioid microphones give fairly high-quality sound in adverse studio conditions (rooms that are too small and too reverberant) and so are useful for the advanced amateur.

At the other end of the scale the supercardioid response is widely used in very high-quality stereo music balance – in the form of a 'crossed supercardioid'. In this, two switchable microphones are placed together but with their axes pointing outward at about 120°.

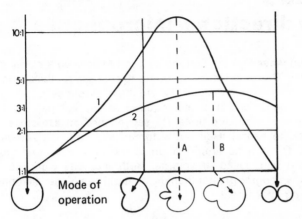

DIRECTIVITY as a microphone ranges from pure pressure operation
(omnidirectional) to pure pressure gradient (bi-directional). Curve 1 shows the ratio
of sound accepted from the front and rear. The most unidirectional response is at A,
for which the microphone is, essentially, dead on one side and live on the other.
Curve 2 shows discrimination against ambient sound. With the supercardioid B,
reverberation is at a minimum. In the range between omnidirectional and cardioid,
direct sound from the front and sides is favoured, so a microphone of this type can
be placed close to a broad source.

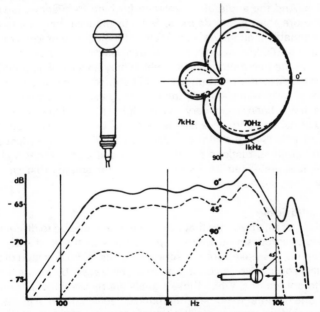

DOUBLE-RIBBON SUPERCARDIOID MICROPHONE, length six inches. Note
that the bass response falls rapidly below 200Hz: this compensates for any
tendency for the bass to rise in close working. But in fact the microphone
would not be used for sound where the lower bass is important, and in
multi-microphone balances it will discriminate against other sources whose
bass radiation is strong.

Highly directional microphones

There are two ways that are commonly used to achieve a highly directional response. One is to focus the sound: to concentrate the sound from
the direction of interest at the point where the microphone is placed. This
results in the amplification of the required sound, compared with unwanted noise from other directions. The second method is to cause sound
from the side to cancel itself out automatically before it reaches the
microphone. There is no amplification of the required sound, but unwanted sound is reduced in volume. Both methods work, but both
degrade at wavelengths greater than the physical size of the microphone's
field-modifying attachment.

Focused sound

Sound can be focused with a parabolic reflector. This may be of fibreglass,
metal, or of transparent plastics (to see action through it); it may be
segmented to be folded like a fan for easy transport or storage.

In practice a reflector that is three feet in diameter can give a gain of
about 20dB for distant sounds along its axis. For faint sounds this results
in a significant improvement in the ratio of signal to microphone and
amplifier hiss. And the angle of acceptance for high frequencies is very
narrow: no more than a few degrees. But the practical limitations are
severe: a reasonably manoeuvrable reflector of 3-4ft diameter loses much
of its effect below 1000Hz. wavefAt low frequencies the response degenerates to those of the microphone itself – which may be a cardioid pointing
the wrong way. Also the microphone is still subject to close unwanted
noises – though where these are in the bass they can often be cut.

Reflectors have been used very successfully for wild-life recording
(particularly birdsong), and, because they work as well indoors as out, for
distant sound effects in the film or television studio (e.g. the dancers'
footsteps and clothing rustle in ballet). But, in general, the sheer bulk of
the reflector has prevented its widespread use for sounds that can be
captured by other methods.

Phase cancellation

In early designs of microphones that used phase cancellation to discriminate against unwanted sound there were bundles of tubes of different
lengths, sampling the wavefront at a series of points in the direction of the
desired sound source. Noise coming at an angle travels by a number of
paths of different lengths. When these signals are recombined in front of
the microphone diaphragm some of them have been changed in phase by
the extra distance travelled, and will therefore tend to cancel those which
have taken the shorter path.

Later it was discovered that a long single tube with a slot in it could be
made to work as well: this is the principle now employed in the gun
microphone.

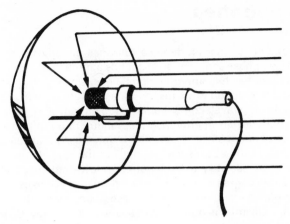

PARABOLIC REFLECTOR   High-frequency sound is concentrated at the focus. The contribution from the rim is important, so the microphone should not be too directional. For low-frequency sounds of wavelengths greater than the diameter of the bowl the directional qualities degenerate.

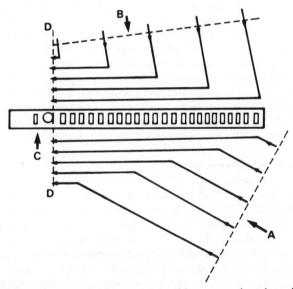

PHASE CANCELLATION   Wave A approaching a gun microphone from a direction close to the axis reaches the front of diaphragm D by a range of paths which differ only slightly in length. There is cancellation only at very high frequencies. Wave B reaches the diaphragm by a much wider range of path-lengths. Cancellation is severe at middle and high frequencies. Normal sound reaches the back of the diaphragm via port C. The polar diagram of the capsule without the acoustic delay network (the interference tube) is cardioid; with the tube it becomes highly directional.

Gun microphones

The mode of operation of the interference tube has already been described: it depends on phase cancellation of sounds approaching from the side. It is a method that has two important limitations. One is common to all highly directional microphones: its response degenerates to that of the microphone capsule at wavelengths below its own physical size. A shotgun microphone with an interference tube about 16in long has an angle of acceptance of about 50–60° from 1500Hz upwards. But at 250Hz it is substantially cardioid (this being the response of the capsule without the tube). It is therefore sensitive to such noises as low-frequency traffic rumble, although some degree of bass cut may reduce this. When used for effects – e.g. a tap dance – or the sounds that accompany most sports (to which it adds great realism) it is often best to use bass cut at, say, 300Hz.

Effect of reverberation

The second limitation is that it is not directional for reverberation. This is because reverberant sound arrives at the microphone by many different paths and so the phase cancellation effect does not work. In live interiors a comparable directional response with better quality might be expected from a good cardioid or supercardioid microphone. A television or film studio may be dead enough for this not to matter too much, though even here it is better to use it only where the subject always faces forward – e.g. in some discussion programmes, but not usually in drama.

When it is used in the open air a windshield is necessary: this is bulky and careful co-operation between sound and cameraman is necessary to keep it out of frame (the best position for good sound is often *just* out of frame). No windshield is necessary when it is on a studio boom (if the microphone is not moved around too fast). This is just as well, because such a large object could cause shadows even from soft-fill lighting.

Pinpointing voices in an audience

One special use is for pinpointing a single voice in a large group such as a studio audience, where there is already general coverage with a cardioid microphone, to which high-frequency presence is added selectively by the gun microphone. For this it is better that the studio should be fairly dead (the audience itself helps to achieve this) and not too big.

For a very large indoor audience that includes potential speakers spread over a wide area, longer interference-tube microphones have been used, for example one that measures 6ft 8in overall. The voices are heard clearly enough, but the quality has not been high, and the change of quality as the microphone swings abruptly to a subject is as distracting as a late fade in.

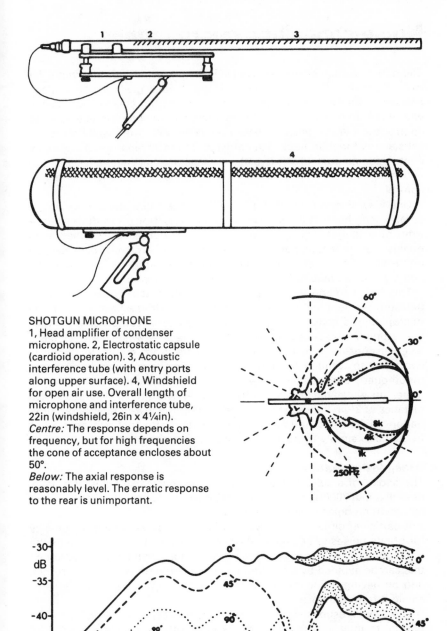

SHOTGUN MICROPHONE
1, Head amplifier of condenser microphone. 2, Electrostatic capsule (cardioid operation). 3, Acoustic interference tube (with entry ports along upper surface). 4, Windshield for open air use. Overall length of microphone and interference tube, 22in (windshield, 26in x 4¼in).
Centre: The response depends on frequency, but for high frequencies the cone of acceptance encloses about 50°.
Below: The axial response is reasonably level. The erratic response to the rear is unimportant.

Noise-cancelling microphones

Microphones often have to be used in noisy places: at sporting events, for example, where (for radio) the time of loudest noise may coincide with the greatest need for explanation by the commentator. One way of dealing with this is to give him an omnidirectional microphone and allow him to control the ratio of voice to background himself. The results are often satisfactory for short items – in particular for news reports – but for long, continuous commentary it is better to separate voice and effects and to allow them to be mixed electrically as part of the job of sound balance and control.

One way of doing this is to put the reporter into a soundproof box, a commentary booth. But this also has limitations. Noise – even whispers – from other people speaking nearby becomes obtrusive, and the sound quality due to the dimensional resonances of a small booth can be unpleasant. So a useful alternative is the noise-cancelling microphone – and a good example of this is the lip-ribbon microphone.

This has a ribbon very close to the mouth of the speaker, and as a pressure-gradient microphone it is subject to extreme bass tip-up. If the microphone is given a reduced bass response to compensate for this, background noise is heavily attenuated in the bass, while the voice has a level response – provided that the distance from mouth to ribbon is controlled to within a fraction of an inch. This is achieved by having a mouthguard, fixed to the microphone, which must be touched against the upper lip when the microphone is in use. The model decribed has a distance of 2⅛in between mouth and ribbon. The low-frequency discrimination against background noise is additional to that achieved at *all* frequencies simply by having the microphone close to the mouth.

Other design features

The choice of a suitable high-frequency response also helps: if it has a peak at about 7000Hz and then tails off, it will cover speech adequately but cut down on noise at higher frequencies.

A particular difficulty that always arises in close working is a tendency for breath noises to be noticeable, and for 'p' and 'b' sounds to 'pop' the microphone (the ribbon is particularly susceptible to this). In the lip ribbon this can be avoided by a design that cups the ribbon *behind* the magnet, and by having a fine-mesh breath screen on top of the microphone to shield it from the nose.

The overall response of such a microphone is generally designed for the normal to loud speech which mixes well with background noise. With quiet speech it may sound a little bassy; if so, it will require bass cut, for example when it is used for very quiet speech in a public concert hall.

The appearance of this microphone makes it unsuitable for use in vision.

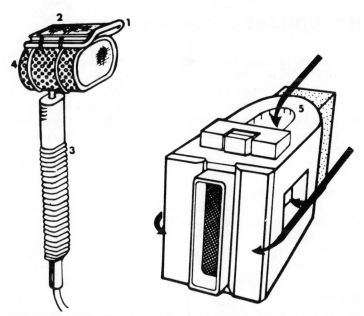

LIP-RIBBON MICROPHONE   Size: eight inches including handgrip. The mouth guard 1 is placed against the upper lip and the stainless steel mesh 2 acts as a windshield below the nose. The handgrip 3 contains a transformer. Within the housing 4 the microphone assembly is mounted with the yoke of the magnet 5 towards the mouth. The sound paths are marked.

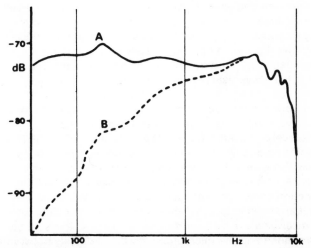

RESPONSE OF LIP-RIBBON MICROPHONE   A, Response to spherical wave at approximately 2½in (i.e. the user's voice). B, Response to plane wave (ambient sound) on axis. This effect is achieved by equalisation, without which B would be flat and A would be very high at low frequencies, owing to extreme bass tip-up at this working distance.

Microphones for use in vision

Everything within a television or film picture, as much as everything in the sound, adds to the total effect. Accordingly, some modern microphones have been designed specifically for good appearance as well as high-quality performance; others have simply evolved into a suitable shape. Indeed, one of the criteria for a microphone is that it shall be reasonably small and neat, so that it does not unduly disturb the high-frequency component of the sound field it is sampling. It should normally be mid- or light-toned, perhaps silver-grey, with a matt finish to diffuse reflection of studio lights.

The condenser microphone lends itself to suitable visual design. Characteristically, a pencil shape about ¾in in diameter has room for a diaphragm about ½in across, usually arranged in the end-fire configuration, with a head amplifier inside the tube behind it. A moving coil unit can also be fitted into the same shape; though with so small a diaphragm area sensitivity may be lower.

It will usually look better coming into frame from the bottom, rather than the top.

Floor and table mountings

The problem of stands is more difficult than that of microphones. The evolution of their design has not carried them so easily towards unobtrusiveness. A floor stand must be solid, heavy and with a sizeable base-area to ensure that the microphone it carries is not knocked over; it must cushion the microphone against shocks and may also have to be telescopic. The design problem extends to connectors and cables, to the knobs for securing extension fittings, and to the clip which attaches microphone to mounting (the clip should be neat but easy for the performer to release in vision; see page 63). But whatever else can be achieved in a long shot there is going to be a vertical line in the picture – so the cleaner this is the better.

Table microphones may sometimes be partly concealed by designing the table to have a sunken well for the stand, but the capsule itself should be above the level of the table top, and should be away from solid surfaces which will form a standing wave at high frequencies. The well should not be fully boxed in or it will generate its own resonances (see page 71). Alternatively, a neat stand and microphone may be acceptable in full vision. Sometimes microphones have been concealed in set dressing – e.g. in flower arrangements. In an ideal world this will be done only if the flowers themselves serve an important visual purpose in a suitable place for the microphone and the microphone itself is definitely not wanted in vision. But on a heavily cluttered desk or table it may be better to hide the microphone behind one of the objects, rather than to add an inconsistent element to the scene.

Microphones with slender, swan-neck extensions may also be useful.

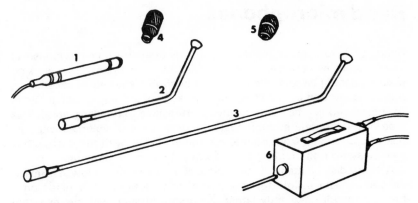

CONDENSER MICROPHONE KIT WITH 'SWAN-NECK' EXTENSIONS FOR USE IN VISION and alternative capsules for cardioid and omnidirectional operation. 1, Head amplifier and capsule (5in). 2,3, Extension pieces which may be fitted between head amplifier and capsule (1ft and 2ft-3in). 4,5, Windshields. 6, Power supply unit.

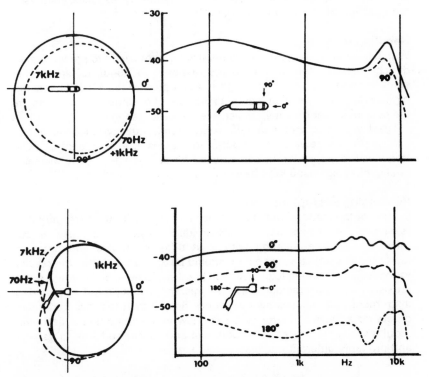

CONDENSER MICROPHONE   Two selections from the kit. Response curves for, *above,* omnidirectional head fitted directly to head amplifier (i.e. without the extension pieces); *below,* cardioid head fitted together with extension piece.

Hand microphones

Hand (or baton) microphones require an omnidirectional response at low frequencies because (unlike the lip-ribbon microphone) the working distance cannot be controlled to an inch or so with any degree of certainty. A moving coil unit can be fitted into a stick about 6in long by about ¾in in diameter, and can be made relatively insensitive to handling noise. This type of microphone accommodates a wide range of volumes with low distortion, ranging from normal use (quiet speech with the microphone well away from the mouth) to loud song at a fraction of an inch.

A minor disadvantage of the moving coil is its relatively low sensitivity. To compensate a little for this, some older designs have relied on a broader tip with a slightly larger diaphragm (though with the expected degradation in high-frequency response); while for some pop singers close working and high sound volume may permit a design with a small diaphragm and its attendant low sensitivity.

There are now condenser microphones which challenge the moving coil in general hand-held work, and exceed it in frequency response. For these, the question of sensitivity does not arise.

Handling techniques
With hand-held microphones it is not always convenient to point the axis directly towards the mouth, so the response at about 45° should be smooth and fairly level – although a broad peak in the 2000–3000Hz (or higher) range may lend 'presence' to the voice. For singers, an equaliser in the control circuit may help to achieve the best response.

There will usually be some lift in the high-frequency response as the axis is turned toward the mouth: this gives the user control over frequency response which can be used for artistic effect, or to increase intelligibility against a loud background.

Rehearsing microphone use
In view of the responsibility for balance that rests with the performer it is reasonable to expect that he should take lessons in the use of the microphone, experimenting with it at different angles and different distances in practice sessions. He should be allowed to judge the results for himself by listening to playback on good equipment. In this way a good musician will learn rapidly to use the microphone as part of a combined instrument that includes his own voice: he will treat the microphone as others do a musical instrument. Each new microphone that is used will require the same process of experiment and accommodation to its new and probably different qualities.

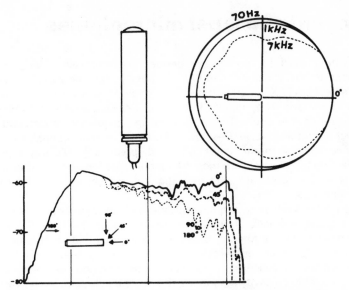

OMNIDIRECTIONAL MICROPHONE WITH SMALL DIAPHRAGM suitable for pop-singer. Length 4¾in. The on-axis response varies within reasonable limits over a wide frequency range. It is robust, with low sensitivity to handling noise. The small size of the diaphragm is compensated by close working.

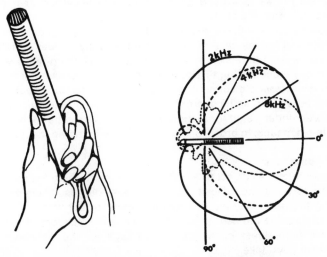

CONDENSER CARDIOID/PHASE CANCELLATION MICROPHONE, overall length: 10in. At 2000Hz and below, the response is cardioid. The minimum distance from capsule to mouth is controlled by the length of the tube, so that there is no significant bass tip-up in hand held use. Phase cancellation due to the interference tube makes the microphone highly directional only at high frequencies. It has excellent separation from surrounding noise for both solo singers and reporters. It can also be set on a table stand or used (often with a windshield) as a short gun microphone.

Lanyard and lapel microphones

Personal lanyard (or neck) microphones are also called 'lavaliers' — pendant jewellery was worn in this manner by Madame de la Vallière, a mistress of Louis XIV. Lanyard and lapel microphones work well in fairly noisy conditions. They benefit from good reflected sound, but are also satisfactory out of doors. They should be light in weight, robust, and not subject to clothing noise. Response should be omnidirectional, because distance from the mouth is not completely predictable. Moving coils are satisfactory but electrets are most frequently used today. They offer good quality at low cost, are extremely light in weight and are relatively inconspicuous.

The sound received by the microphone is affected by its position close to the chest, the lack of high frequencies under the chin, the room acoustics and by the filtering effect of any thick clothing that is used to conceal it. The required response varies, but in general some top lift (2500–8000Hz) is necessary and in some older (moving coil) designs has been generated by cavity resonance above the diaphragm. In one version a circular ring or clip can be raised around the head of the microphone, an arrangement that has the advantage of allowing the microphone response to be tailored acoustically to the needs of other situations without having to use equalisation in the microphone channel. Also it can readily be converted to use as a hand microphone or in television or films to be concealed within the shot.

Madame de la Vallière made a point of having her pendant clearly visible — and for the sake of good sound we may do well to follow her example: even a thin layer of silk may produce clothing rustle while soft wool may reduce the high-frequency response. A pendant microphone in vision is neater if the colour of the cord matches its background.

If it is essential that the microphone is not seen, an electret can be taped to the skin under a shirt, or to the clothing itself. It may be advisable to use double-sided tape, in order to avoid rubbing or bumping the microphone.

Methods of use
The user must be subject to a certain discipline: turning the shoulders or leaning forward may cause clothing rustle, bumps or crackles, while turning the head may put the speaker off-mike. But within the limits of what is possible the user should be given some freedom of movement, both for naturalness of picture and to cater for the unexpected in an otherwise good take.

The cable should be concealed within a jacket, say, if the widest shot is to the waist; or run down inside a trouser leg for a wider shot. If the subject is to walk about, use a radio link, or if this is impossible tie the cable into a shoe lace or tape it to the shoe. There should be a connector somewhere near at hand and easily accessible so that the user can be unhooked temporarily when necessary.

58

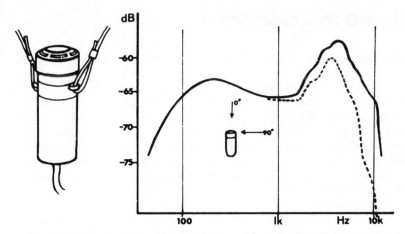

LANYARD MICROPHONE   Omnidirectional moving coil. Length 2½in, weight 2¼oz plus cable. The response curve compensates for the distortion of the sound field that is found in this position.

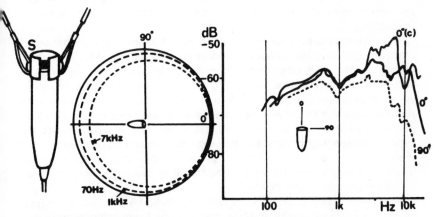

LANYARD MICROPHONE with adjustable sleeve or slip, S, around diaphragm. Omnidirectional moving coil. Length 3in, weight 1¼oz plus cable. With the clip raised (curve C) a resonance peak makes it suitable for use as a lavalier microphone. With the clip down, it may be hand held.

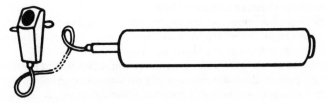

LAPEL MICROPHONE   Length, 1½in. Diaphragm diameter, ¼in. Omnidirectional condenser with remote electronic section which can go in a pocket.

The use of radio links with personal microphones allows greater freedom of movement and better pictures.

Radio microphones

The increasing simplicity and reliability of local radio link equipment for use with personal microphones has led to their use in progressively more marginal situations including studio and film location work which would formerly have relied on cables connecting the performer to the sound mixer. The use of a wireless microphone avoids the ludicrous sight of a cable being dragged around by a performer, or of the cable getting tangled up; and gives him more freedom of movement between rehearsals or takes. The alternatives (perhaps a studio boom or a gun microphone) should always be considered before settling for a lanyard or lapel microphone. Radio links can, of course, also be used with hand-held microphones.

The transmitter may be in a pocket of pouch in the user's clothing: this may require co-operation between sound and wardrobe departments.

Aerial and receiver
In a studio the microphone screening cable may double as a simple quarter-wavelength aerial; alternatively the aerial may trail down a trouser leg.

The receiving dipole must be placed where there is an unobstructed electromagnetic path between performer and aerial. This can 'see' through wood and other nonconductive materials, but is likely to be screened by metal objects – including the large camera and sound dollies which move about a studio. Before a radio microphone is given to a performer in a studio, the sound man and his assistant should check its performance in all the places where it is likely to be used and locate the aerial or aerials accordingly. Existing monitor circuits can be used to carry the signal from dipoles to a receiver in the sound control room. Note that in the studio another radio link is normally in use to provide the floor manager with director's talk-back.

Studios are notoriously difficult places for radio links, often requiring a diversity of pick-up points; outside, links may be technically easier (if less predictable). They can now be used to allow a news reporter to remain completely separate from camera and sound recording equipment, and yet be ready to record at all times.

Batteries, compressors and switches
A radio microphone must be battery-powered – and when it fails the battery is likely to go fairly fast. In crucial uses a log of hours of operation should be kept and the battery replaced in good time.

A compressor-limiter in circuit between the microphone and transmitter (normally as part of the transmitter pack) prevents overloading. But the level fed to it should be preset in rehearsal to avoid noticeable compression effects. In addition a switch should be provided so that the performer can obtain privacy when he needs it.

60

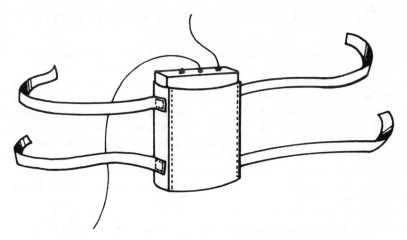

POUCH SUITABLE FOR RADIO-MICROPHONE TRANSMITTER of older type. It should be made exactly to fit the transmitter box and have tapes securely attached; 24 inches on either side, top and bottom. In a pocket the transmitter may jolt against the body, and with this type of equipment the result may be signal bumping. Modern equipment with miniaturised solid state electronics is more robust, and easily fixed inside the clothing.

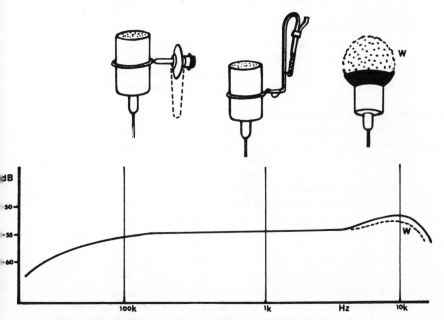

VERY SMALL PERSONAL MICROPHONE   Size: 0.7 × 0.42in. Weight of condenser capsule, 0.16oz plus 10ft of cable permanently attached to supply unit (weight 3.5oz). The microphone can be used with one of several clip attachments or a windshield, W.

61

Other microphones and equipment

Low-quality actuality microphones are sometimes linked to professional radio, television recording equipment. For example, the telephone system may be used as a source in radio, in order to obtain news reports from remote locations from which broadcast-quality lines are not available, or to obtain full audience participation. The telephone transmission band-width is approximately 300–3000Hz. The response of the microphone is not flat and distortion and noise levels are high by broadcasting stan-dards. Nevertheless, the results are usually acceptable on the air if the reason for it is clear; intelligibility is generally adequate.

To get a strong signal with low noise it may be better to avoid a local exchange and to have a special line installed between the studio and some point close to the centre of the area telephone system. A feed from the telephone circuit should be passed through an amplifier and limiter to the control desk, where it will be mixed with studio-quality sound. Ensure that the remote speaker will be talking directly into the mouthpiece before switching on-air.

As quality falls below that normally expected on the telephone, the audience's tolerance diminishes rapidly. For example, intercom between a pilot and control tower is usually difficult to make out by the unpractised listener, often because of high background noise levels.

Underwater sound

To pick up underwater sound, enclose and seal a stick microphone within a protective rubber sheath and suspend it in the water by it own cable. Note that under the sea an omnidirectional microphone is likely to pick up a great deal of noise from seawash, ships or passing aircraft.

Microphone mountings, etc

Whether in the studio or on location, mountings must serve a number of purposes. They must set up the microphone in the right place and pointing in the right direction; and often also allow freedom for man-oeuvre. They must ensure that a microphone that is put, stays put, and they must not transmit to it any rumble or vibration from studio structure or furniture that is capable of being picked up by the microphone. Some mountings that are found in a radio studio are shown opposite. As integral parts of microphones or their balance, they are also shown throughout the text.

A wide variety of plugs and sockets are available for connecting microphones to power units, mixers and recorders. We inherit the results of a multiplicity of standards and individual designs. As a result of this – and also because of the range of impedances that are used for different purposes – the ready interchangeability of equipment cannot be assumed. Except where items are standard within a studio or organisation, adap-tors, transformers and perhaps even a soldering iron may be necessary.

EASY-RELEASE CLIP on floor stand allows performer to take the microphone from the stand while in vision.

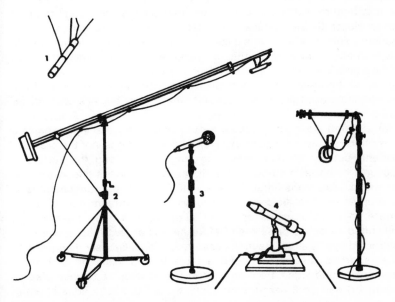

MICROPHONE MOUNTINGS   1, Suspension by adjustable wires. This arrangement gives complete control over position and angle, but sometimes it is sufficient to suspend the microphone by its own cable. 2, Boom, adjustable for angle and length of arm, as used in radio or sound recording work. 3, Floor stand (with telescopic column). 4, Table stand (but 1, 2 or 5 are better for table work if there is any risk of table-tapping by inexperienced speakers). 5, Floor stand with bent arm. All methods of mounting have rubber shock-absorbers or straps.

An acoustically transparent outer shell can protect the microphone with little effect on its response.

Windshields

Windshields are supplied to fit most microphones. Sometimes they may even be needed indoors to reduce breath effects. Generally, however, their purpose is to cut down wind turbulence at sharp or low-radius edges and prevent its penetration to the region of the diaphragm. To reduce wind effects to a practical minimum, a smoother airflow is obtained by increasing the radius of the curves over which the air travels: windshields are generally made up of relatively large spherical and cylindrical sections. Windshields do make a slight difference to the frequency response, but usually within the same tolerances as the microphone without the windshield.

Microphone mice

Microphones can be placed unobtrusively in vision in a wide shot by inserting them within a foamed rubber or plastics pad that acts as a combined windshield and shock-absorbent mount, and simply putting them on the floor.

Generally, microphones are kept away from reflective surfaces owing to the interference fields these produce. However, as the diaphragm is moved closer to the surface, this effect moves up through the sound frequency spectrum, until at a distance of 1in it occurs only in the highest octave of human hearing, so that on speech and song the effect becomes unimportant. As a result, if the surface is solid enough to reflect sound well at all frequencies, a good-quality balance can be obtained by a microphone placed within an inch of it. The type of balance depends on the microphone's operating principle.

For a sound wave approaching a rigid solid surface at right angles a microphone so placed registers only pressure variations, not the pressure gradient, but for a sound travelling lengthways along the surface it measures by either method. A directional microphone is effective therefore only in directions parallel to the wall. A common arrangement is to place an end-fire cardioid microphone with its diaphragm normal to the surface, directed towards action at a distance. It may therefore be well-suited to opera performed before an audience. Sightlines are not obstructed by the small, neutrally coloured, and felicitously named mice.

For a group discussion round a table it is possible to pick up a reasonably good signal from an electret placed within a foamed plastics (or rubber) pad on a hard-topped table. Since the microphone is omnidirectional, the speakers may sit on all sides, but fairly close, and the acoustic should be rather dead.

The PZM (pressure-zone microphone) is specially designed to use the same boundary principle: it is placed flat on (or against) a solid surface.

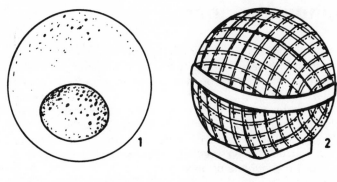

WINDSHIELDS 1, Open cell plastic foam. 2, Glass fibre and polyester screen. Typically the diameter of these may be 2in or more.

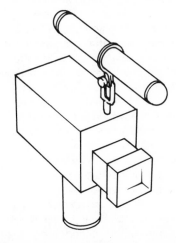

GUN MICROPHONE ON TELEVISION CAMERA A gun microphone may be used in a protective windshield to cover sports or other events, pointing toward the same action as the zoom lens.

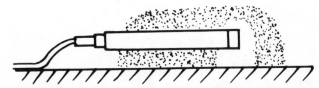

MICROPHONE MOUSE, in section. It is made of foamed rubber or plastics, moulded with a tubular slot for the insertion of a microphone and a cavity for the capsule. It is placed on the floor.

Microphone balance

The most modest requirement of any microphone in use is that it must convey sufficient information – whether this be in terms of the intelligibility of speech, or the content of musical sound or effects. When a microphone is used to pick up speech in difficult acoustics or noisy conditions, this need may override all others. But where more than one limiting position of the microphone is possible, there is a series of choices that we may make. The exercise of these choices, the selection and placing of the microphone in relation to the source and the studio acoustics, is called microphone balance.

The objectives of balance
The first objective of microphone balance is purely technical: to pick up the required sound at a level suited to the microphone and recording set-up, i.e. to convert the acoustic energy of sound to a corresponding electrical signal with a minimum of irreversible distortion. The next objective is to discriminate against unwanted noises.

In professional practice this means that we must decide whether to use one microphone or several, and select those with suitable directional and frequency characteristics. In many cases we may also be able to arrange the position of the source or layout of several sources.

These decisions depend on the next and, some would say, most important aim of a good balance, which is to place each microphone at a distance and angle which produce an aesthetically satisfying degree of reinforcement from the acoustics of the studio. That entails controlling the ratio of direct to indirect sound picked up by the microphone, and making sure that the indirect sound is of a suitable quality.

Acoustic reinforcement may, however, be replaced by artificial reverberation which is mixed into the final sound electrically. In this case each individual sound source can be individually treated. For this reason, and because picking up a sound on a second more distant microphone reintroduces the effect of the studio acoustics, an alternative aim of microphone balance is to separate the sound sources so that they can be treated individually. For the man working on pop music the term balance means a great deal more than microphone placement. It includes treatment, mixing, and control, all integral parts of the creation of his single composite sound.

Balance tests
Good balance can only be judged subjectively, and the best way of achieving it is by making direct comparative tests in good listening conditions between two (or more) microphones and moving one at a time until the best sound is achieved.

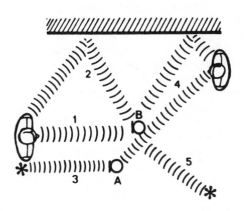

BALANCE TEST: Compare the sound from microphones A and B by switching between the two. The microphones lie in a complex sound field comprising many individual components.

Listen for:
1. Clarity and quality of direct sound.
2. Ratio of direct to indirect sound. Does this add body, without being obtrusive or heavy (or otherwise unpleasantly coloured)?
3. Unwanted sound from the source or close to it. Can this be prevented or discriminated against?
4. The relative volume of other (required) sources, direct or indirect.
5. Unwanted sound from other sources.

The variables: Change only one of these, or as few as possible, between successive comparison tests.
i.   Distance from the source.
ii. Height and lateral position (at a given distance).
iii. Angle of directional microphone.
iv. Directional characteristics of microphone (either by internal switching or by changing the microphone).
v.   Frequency response (by changing the microphone or its equalisation).
vi. The use of additional microphones.

If sufficient microphones are available for adequate balance tests, trial recordings may be made, and replayed one after the other, 20 seconds or so at a time.

Speech balance

In a good speech balance, the voice is clear and natural-sounding, perhaps reinforced to some degree by the room acoustic. Microphone type and distance are important. A deep voice is over-emphasised by close working on a microphone that is directional at low frequencies and therefore subject to bass tip-up. Electronic correction (in the microphone channel) may be used to compensate for this, and does not affect voices that are less bassy, but it does limit movement backward and forward. The distance between speaker and microphone must remain constant or the bass frequency content will vary, and the balance between direct sound and reverberation (in a live studio) will see-saw – especially if movement is sufficient to warrant the control of volume.

A sibilant voice is made worse by a microphone with an erratic high-frequency response. This, too, can be corrected electronically, but only at the cost of introducing 'holes' in the top. A microphone with a smooth response in the first place would be better. It is not essential that this response should be extended into the extreme high-frequency range: high-quality microphones are not usually necessary for speech.

For many purposes, some acoustic reinforcement is good. Smaller studios are, however, likely to introduce coloration which can be reduced only by electronic filtering or closer working. Film commentary usually requires a fairly dead acoustic, in order to avoid conflict with picture, and to separate narrative from action.

Directional microphone

A directional microphone is generally used for speech. A response which is cardioid over most of its working range, but which degenerates to omnidirectional in the extreme bass, can be used by a single speaker at 12in or less. Unless the studio is unusually dead an omnidirectional microphone has to be used much closer so that balance between voice and acoustics is impracticable: such a microphone is therefore more likely to be used only for very close working, to eliminate acoustics and unwanted noise as much as possible (though for this a directional microphone might still be better). A typical application might be for the voice of a disc jockey who will be heard on low-quality or car radios, or against noisy listening conditions.

Bi-directional microphones (ribbons) have been widely used in radio studios, even for single voices. Unless corrected for close working, balance is at a minimum of 18in to 2ft, depending on the bass content of the voice. This gives a full account of the studio acoustics (even though pick up is limited by the relatively narrow live angles) and these must therefore be of good quality, i.e. without marked coloration. Small studios are not generally suitable for this type of balance.

INTERVIEWING TECHNIQUES USING DIRECTIONAL MICROPHONE (e.g. supercardioid)

Position 1: Microphone at waist level: sound quality poor to fair. In television, this position may sometimes be used in order to clear the picture, but only if background noises are low, and the acoustic dead.

Position 2: Microphone at shoulder level, static. Sound quality fair to good, provided that both speakers are close together – e.g. standing close and at 90° to each other, or sitting side by side.

Position 3: Moving the microphone from one person to the other. Sound quality good; but the aggressive use of the microphone may either distract the interviewee or allow him to move closer in to it than is desirable.

An omnidirectional microphone would need to be close; it would give very poor sound in the waist position. Also consider using a gun or neck microphone when out of doors, or a neck or boom microphone when in the studio or an indoor location.

Two or more voices

Speech balance for two voices can be accomplished by using separate microphones for each voice, or a single static microphone for the two; or by a moving microphone which is directed to each speaker in turn. Further voices may be covered by additional microphones for individuals or pairs; by balancing more voices to a single static microphone; or, again, by directing a movable microphone to individual speakers or groups of speakers in turn.

Using separate microphones
As more and more microphones are used, the effect is to open up progressively more of the studio acoustics. If many microphones are open at the same time, closer balance for the individual microphones is necessary, or a deader acoustic is required than if only one microphone is used. In principle a more sharply directional response from each microphone would also do the trick, but as we have seen, there are fundamental limits to the directional qualities of small microphones (and even gun microphones degenerate to cardioid in lively acoustics). The alternative is to hold back all microphones except that for the main speaker at any one time: this also tends to improve intelligibility when several people are speaking together. It is not necessary to fade down the other channels all the way; 6dB may be sufficient. In an informal discussion, the sound mixer needs to see all of the speakers directly all of the time, in order to anticipate each contribution. Where a wide area is to be covered and there is no direct line of sight, put a sub-mixer close to the action.

Unidirectional microphones may be used for pairs of speakers – e.g. in a panel game. Sometimes, however, one speaker may spend much of his time talking away from the microphone. If this happens, the sound man must make up his mind whether that is worse than having an extra microphone to control.

Advantages of bi-directional microphone
In radio there is still much to be said for the bi-directional microphone which accommodates up to four speakers easily (two sitting on either side at distances of about two feet) or six at a pinch. The position of each speaker may sometimes be adjusted a little for his natural volume and voice quality, so that minimal control may be exercised by one sound man (who again should be able to see the speakers).

A group discussion with, say, six speakers can be balanced on a cardioid above the centre of the group and directed downwards – or below, directed up. This is not an ideal arrangement because the volume will be set for off-axis speech and the microphone is therefore particularly sensitive to noise and reflected sound in the line of the main axis.

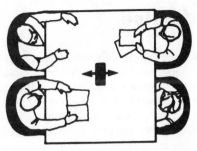

BI-DIRECTIONAL MICROPHONE (usually a ribbon) in layout for studio discussion with up to four speakers.

PANEL GAME BALANCED WITH TWO BI-DIRECTIONAL
MICROPHONES   Each microphone is arranged so that the other pair of speakers is on its dead side. But any sound from in front of the desks (from the audience, for example) will be picked up on both: the microphones must therefore be in phase to signals from that direction. Supercardioid microphones are likely to be equally satisfactory.

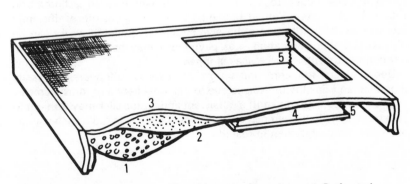

ACOUSTIC TABLE WITH WELL   The table top has three layers. 1, Perforated steel sheet. 2, Felt. 3, Woven acoustically transparent covering. 4, Wooden microphone platform, suspended at the corners by rubber straps, 5.

71

Background noise

In all natural locations there is some normal level of background noise or 'atmosphere'. It is generally acceptable in films, radio and television reporting, etc, provided that there are no obtrusive, inappropriate, or very loud elements. Examples of the most inconvenient types of noise encountered in filming are passing aircraft or unseen traffic. It is difficult to discriminate against, even with directional microphones; the best safeguard is a suitable choice of location.

Apart from such difficulties, low levels of background sound in a location are good: they help to establish or confirm the character of the place. Whether on location or in the studio an 'atmos' or 'buzz' track should be recorded at the same level as the master sound takes to help the editor avoid unnaturally quiet gaps between edited passages.

Studio noise problems

In a studio, noticeable background noise has no natural place, except in a play where the atmosphere of a 'real' location is to be recreated, in which case it is the deliberately-composed sound effect and certainly not studio noise that is required.

Ventilator hum can sometimes be obtrusive, particularly with quiet voices. Comparative tests with various volumes of voice and with different microphone layouts will show the limitations of a studio and how to keep such noise to a minimum. Plainly, this problem can be minimised by using a close balance and a strong voice. But this, in turn, is a limitation on use both of the natural (or designed) acoustics of the studio and of the quality of speech (or type of speaker) that it is being used for. You may not wish to have an unpractised but naturally quiet speaker change his whole style of delivery for purely technical reasons (though this would be better than his not being heard).

Structure-borne noise is another nuisance. Some studios have complex forms of construction to avoid this (e.g. the whole room is floated on rubber blocks). Steel-framed buildings transmit sound more efficiently (i.e. worse) than older, more massive structures. Building works can be obtrusive. Polite persuasion – or payment – may be necessary to get hammering or other loud noises stopped.

Doors, chairs, footsteps and scripts, all potential offenders, are easily identified and dealt with. Noises due to personal habits or idiosyncracies may be more difficult to spot: a persistent and erratic click may turn out to be a retractable ball-point pen or a loose dental plate. In radio such sounds are more noticeable than in real life.

MICROPHONE POSITION   In a good
microphone position for speech the
head is held well up and the script a
little to one side. The speaker should
work to the microphone and not to the
script or down towards the table. The
script must never be allowed to drift
between mouth and microphone.

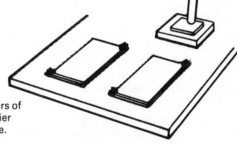

AVOIDING SCRIPT NOISE   Corners of
script are turned up to make it easier
to lift pages noiselessly to one side.
Stiff paper is used.

SCRIPT RACK   The angle of the script
prevents sound reflections from
reaching the microphone. The speaker
should avoid dropping his head as he
reads to the bottom of the page. The
rack (as well as table surface) may be
made of acoustically transparent
material. The slung microphone will
discriminate against table-borne
noise.

Speech balance can be changed creatively by altering the acoustic furniture surrounding speaker and microphone.

Drama studio acoustics

Studios with a range of acoustics are still used for radio drama in countries where this survives. These studios have two or three separate areas with different qualities of sound reinforcement. These may be bright ('live'), normal or dead; and they may be varied further by the use of heavy drapes and movable screens. Truly dead acoustics – representing the open air – are difficult to simulate realistically within the studio.

One area – the deader end of the main studio – will, in fact, differ little from a normal indoor acoustic, being deader only by enough to permit normal working distances of 30in to 4ft with a bi-directional microphone (ribbons remain very satisfactory for monophonic drama). This gives a realistic simulation of the actual conditions it is supposed to represent, as does the live area (which could also be used for a small music group). But note that a space with lively (and therefore dominant) acoustics has certain built-in limitations. Its size may be registered audibly by its dimensional resonances. A small bathroom and a large hallway may have the same reverberation times but they are recognisably different. The characteristics of the reflecting surfaces (e.g. wood panelling) may also be recognisable, and noticeably different from those of, say, a hall lined with marble. The confined space within a car can be simulated by the use of padded screens.

Action and narrative
If action is to be combined with narrative a separate microphone will be provided. In the past this has been a ribbon, near a confining or reflecting surface, and often with bass equalisation for close working; or an omnidirectional microphone used close. Better, however, may be a cardioid or similar microphone, still used fairly close: characteristically the narrator should appear to be closer than the action, and with less acoustic 'dressing'.

Ideally the narrator should be well separated from the action – but if he is a participant in it, it is more convenient for him to have his own microphone close to the main microphone, in which case he will either need screens or his microphone should have a characteristically different frequency response. As different designs fall short of the ideal of a level frequency response in a variety of ways, that should not be too difficult.

The layout described here is suitable for continuous recording – a technique that is still preferred by many. But with stop and start recording, the acoustics can be modified in the gaps between takes.

74

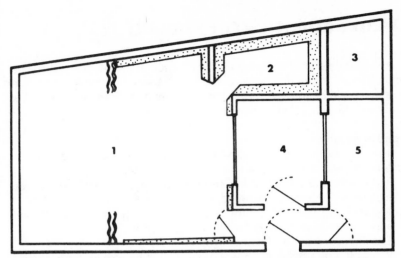

DRAMA STUDIO SUITE   Many such studios still exist. The areas shown are:
1, Main acting area: the (relatively) live and dead ends of the studio can be
partially isolated from each other (and their acoustics modified) by drawing
double curtains across. 2, Dead room, with thick absorbers on walls. 3, Echo
chamber (or equipment room, including space for reverberation plate). 4,
Control cubicle. 5, Recording room. In this example the irregular shape of the
site is an advantage: the use of non-parallel walls avoids standing wave
coloration.

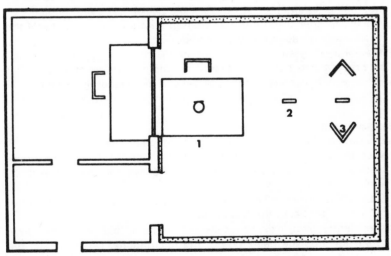

SIMPLE LAYOUT FOR DRAMA   A simple set-up using a cardioid
microphone, 1, for narrator, near the window, and two ribbons in the open
studio for the actors. Microphone 2 is 'open' for normal indoor acoustics and
microphone 3 is enclosed in a 'tent' of screens to represent outdoor quality.
This is a less versatile layout than is possible in a studio that has a built-in
range of acoustics but it may be adequate for low budget productions with
small casts, or where the broadcast is likely to be heard in poor listening
conditions.

Dead acoustics for speech

Open-air scenes are frequently required in sound drama, but recordings actually made out of doors are subject to extraneous and often unsuitable noise. It is therefore necessary to simulate the open air – with its characteristic lack of reverberation – in the studio.

Dead acoustics in the studio

A truly dead acoustic – as can be obtained by the use of absorbent material a yard or more thick – has certain advantages:

1. It provides the best possible contrast to other acoustics in use, so making a wider range of sound quality possible.
2. The muffling effect of the treatment causes the performer to lift and edge his voice as he really would in the open air.
3. Effects recorded out-of-doors blend in easily.
4. Voices that are made to appear more distant by placing them toward the dead side of the microphone do not generate parasitic studio reverberation to ruin the effect.
5. There is more space than inside a tent of screens.

But there are also disadvantages:

1. A truly dead acoustic is uncomfortably claustrophobic to work in.
2. Completely dead sound is not so easy to balance: the muffled voices must either be lifted (in which case the peaks may overmodulate), or allowed to appear more distant than interior sound. Alternatively the level of interior sound may be held down.
3. It is not so pleasant to listen to for long periods.

The first of these is the main reason for finding an alternative. One that has been widely used is a tent of screens in the deader end of the main studio.

Using screens

With a double-V of screens round a bi-directional microphone:

1. Keep the screens fairly close to the microphone. This restricts movement a little but keeps sound path-lengths between reflections short, and reverberation low.
2. Set each pair of screens to form an acute-angled V. Actors should not retreat too far back into this angle.
3. Keep 'entry' and 'exit' speeches within the V, but physically move round towards the dead side of the microphone. Do not direct the voices out into the open studio.
4. Off-microphone (i.e. distant) speeches may be spoken across the angle of the V, giving the voice less volume but more edge.

Such techniques are half-measures, perhaps, but acceptable as a convention. The main thing is to set up a satisfactorily contrasting range of acoustics.

DEAD ROOM ACOUSTIC TREATMENT Wedges of soft foamed plastics cover solid walls. In a room to be used for acoustic measurements the floor might also be treated similarly, with a false floor (an acoustically transparent mesh) above the wedges.

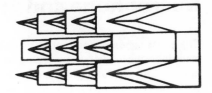

USE OF SCREENS Double reflection of sound in a V of screens set at an acute angle. For any particular path there are some frequencies which are poorly absorbed, but these are different for the various possible paths.

A TENT OF SCREENS If there are parallel surfaces, 1, standing waves may form. It is better that sound should be reflected twice on each side, 2, This breaks up and absorbs more of the sound waves.

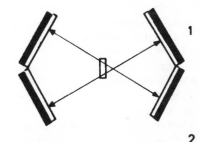

1

2

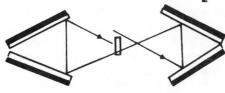

SPEAKING OFF-MICROPHONE No actor should go farther than this, or turn farther into the screens. The effect of greater distance must be achieved by a change of voice quality, or by adjusting the microphone fader.

In television and film the normal criteria for good speech balance still apply but the methods used must be visually acceptable.

Microphone and picture

There are several conventions which may be consulted on the use of microphones in vision. According to one, close pictures should have close sound, while a wider picture may have rather more open (more reverberant) sound. This can often be satisfied by having a movable microphone just out of frame. But for many programmes, in the case of a subject who always remains in the same place (e.g. in a seated discussion, or quiz) such variations in sound quality add little or nothing to the effect, while making lighting more difficult (because of microphone shadows). Also, in very wide shots intelligibility may be reduced – and the need for adequate intelligibility usually overrides the matching of sound to picture. For many purposes it is simpler to accept that microphones may be placed where the camera sees them.

Where chromakey (otherwise known as colour-separation overlay or 'blue-box') is used to provide a setting, it is the background of the composite picture that governs the desired acoustic. Even when the setting is physically present, it may sound different from how it looks, requiring some compensation.

When a microphone may be seen

This is the subject of another convention: they should not be seen in dramatic presentations (including situation comedy) where they would spoil the attempted realism, or get in the way of the action, but may appear in most other programmes, such as for example news, discussions, quizzes or musical entertainments. This is not to say that microphones should be in vision in all of these. Their very ubiquity is a point against them: there is visual relief in getting rid of them when the opportunity allows.

A further convention requires that unless the microphone is carried by one of the performers, it should not be seen to move. If it did, it would be distracting. The breaking of any of these conventions must be intentional, and for a purpose that is clear to the audience.

In films, including television films, the same conventions apply. But here the use of microphones in vision needs to be more discreet, because most films are shot in places where microphones are less a part of the furniture than they are in television studios: seeing the microphone may remind the viewer of the presence of other technical equipment, together with the director and the camera crew. Also, in film there may be greater variety of picture and faster cutting; in these circumstances a microphone that is erratically in and out of vision can be more intrusive.

The use of table, stand, hand and neck microphones in vision has been discussed (see pages 54–59), also the use of gun microphones (pages 48–51). In the studio there are other techniques: notably *the studio boom,* which serves to place a microphone just out of frame (pages 79–83).

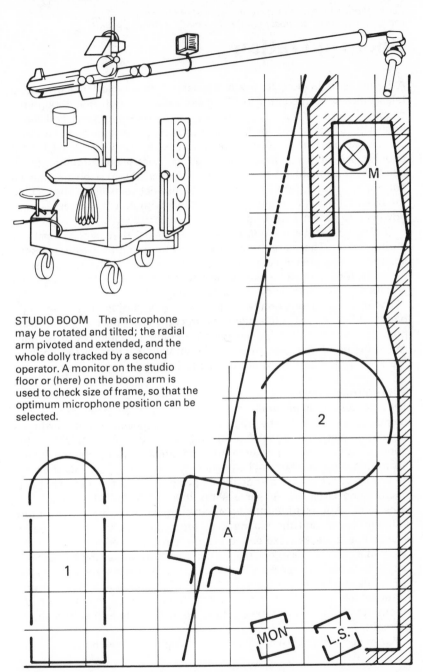

STUDIO BOOM   The microphone may be rotated and tilted; the radial arm pivoted and extended, and the whole dolly tracked by a second operator. A monitor on the studio floor or (here) on the boom arm is used to check size of frame, so that the optimum microphone position can be selected.

MICROPHONE BOOM   On studio plan (actual size, using 1:50 metric grid): A, Boom, indicating shortest and fully extended lengths of arm. 1, Small motorised camera crane. 2, Pedestal camera. MON, Monitor, L.S., Loudspeaker. M, Slung microphone to cover speech in a corner inaccessible to the boom when the actors are separated.

Boom operation

A boom used in both film and television has a typical reach ranging telescopically from about 7 to 17ft. It can swing through 360° (though the shape of the operator's platform does not make this easy) and tilt to angles of 45° upward or downward. The microphone at the end of the arm can be rotated to favour sound from one direction or discriminate against another. All of these movements can be controlled continuously. In addition, the height (of the point about which the arm pivots) can be adjusted over a range of 3ft (in the typical case between 6ft 5in and 9ft 5in), the operator's platform going up and down with it. It is normal practice, however, to preset a working height of about 7ft.

The boom can be tracked (on inflated wheels) by a second operator during a shot. There are three wheels and the pram (or dolly) is readily steerable – though it is a clumsier object to move around than most studio cameras. Moves must be planned in advance so that the line of the pram and its one steerable wheel are suitably aligned.

The operator stands on the platform to the right of the boom, his right hand holding the crank to extend or retract the arm (this crank is near the point of balance). With his left hand he holds a handle which he pulls round the boom to rotate the microphone; with this arm he also controls the angle of the boom – which can be locked off in any position.

Microphone position

The microphone is generally a cardioid, and its ideal position for many scenes is as close to the edge of frame as possible. For maximum separation between signal (voice) and noise, the operator brings the boom in front of each speaker in turn and directs the microphone toward his mouth. During rehearsal he repeatedly drops the boom into the edge of frame, checking on a monitor what his limit is for each shot. His microphone is closer for some shots than others, but his variation is consistent with picture: the tighter picture is accompanied by more intimate sound and the perspectives are right. The sound for exterior scenes has to be as close as the picture permits; interiors may then be allowed a slightly looser balance for contrast. Sometimes the microphone 'splits' between two speakers: in this case each voice is given equal value.

For talk shows requiring greater flexibility of operation and to allow for unrehearsed wide shots a gun microphone may be mounted on the boom: its high directivity means it can be held a little further back than a cardioid.

For music a high-quality condenser microphone may be used.

80

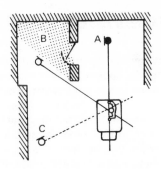

BOOM OPERATOR IN DIFFICULTIES If the operator of a boom positioned for use in area A is immediately asked to swing to area B, he is blind to all action which is obscured by the wall, and so cannot judge the angle and distance of his microphone. If the action continues to area C, again without movement of the dolly, the operator is supported only by his boom and a toe-hold on the platform. Better provision must therefore be made for coverage of B and C before the position for A is decided.

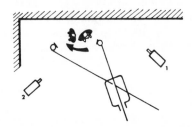

BOOM OPERATOR IN A HURRY A simple movement by the performer may require a complex movement of the boom, here swinging it to the left, extending the arm, and turning the microphone to the right. Dialogue which continues during a rapid move may go off-microphone. How tolerable this is depends on the camera position: if the action continues on Cam 1 it may be acceptable; if there were a cut from Cam 1 to Cam 2 it would be unsatisfactory; if the whole action were taken on Cam 2 it would be ludicrous.

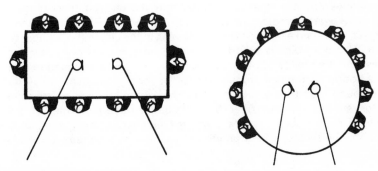

BOOM COVERAGE of a large television discussion (e.g. a scene showing a board meeting in a play).

In television you need good sound coverage that does not get in the way of good pictures.

Sound in the television studio

When a boom is chosen as the main instrument of coverage it requires precise co-operation between its operator and the sound supervisor. The boom operator has headphones with programme sound on one phone and director's talkback on the other.

In the talkback circuit to sound staff on the floor, the sound supervisor can override the director's line (in BBC practice there is a tone-pip as he cuts in on the line). He can warn the operator of changes foreseen on his preview monitor or clear him to a new position at the end of a scene: the sound supervisor can follow a script, while the operator has only a card listing his shots and his own notes on framing. In cases where the operator cannot easily see a floor monitor (which should be as rare as possible) the supervisor will have to talk him through the framing of the shots. Better than a floor monitor, however, is for each sound boom to have a miniature monitor clipped on to the boom arm just in front of the operator.

If the director or cameramen introduce wider shots than have been rehearsed there is a risk of the boom intruding in vision. The alternative is to allow sound to suffer by adopting a looser balance. Good sound demands co-operation and self-discipline all round.

Cables

The sound cable is multicored, to carry the microphone output and the various communication circuits. The front of the boom pram is a good place from which to point a loudspeaker into the set, so this too may be fed by the same route. The sound cable may be dropped from the lighting grid (where several microphone points will be available) at a place where it does not get in the way of lights or cameras. The cameras must be able to move rapidly from one set to another without running over sound cables, and the cables should not be visible in shot.

Additional microphones

Microphones other than that on the boom may also be used; these may be slung in awkward corners, suspended from hand-held 'fishpoles', hidden in the set, or placed as personal microphones in the clothing of or near to actors. Where a voice is heard on two or more microphones in rapid succession, the quality of other microphones should be matched to that on the boom. In complex productions several booms are used.

The sound man tries to accommodate all of the movements that the director requires of a performer. But both director and performer should be aware of the sound man's problems and so consider what may reasonably be accomplished: again, compromise may be required.

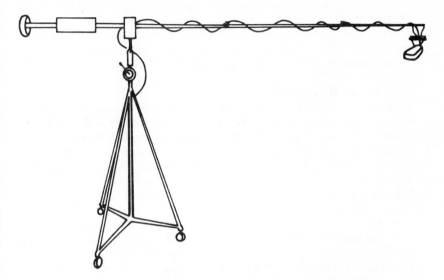

SMALL STUDIO BOOM (about 5ft high) in which the arm can be swung and the microphone angled by gripping the handle at the back of the boom with one hand and the body of the tripod stand with the other. The extension of the arm is preset, but the whole stand can be rolled bodily toward or away from the action.

FISHPOLE   This provides a highly mobile microphone, but is tiring for the operator. Held like this the rod may cross too low and get into the picture of a second camera, or cause a shadow. A second position in which the operator balances the boom above his head is possible but even more tiring.

Booms and lighting

A key light throws a hard-edged shadow. If it falls on the boom the shadow moves as the boom moves. Other lighting includes the fill (a diffused light from the opposite side to fill in the shadows) and rim or back lighting that shines from above and behind a subject to separate it from the background. These cause less trouble with the boom.

Avoiding boom shadow
If the key and boom come into the set from a similar angle, there is danger of a shadow being thrown down on to the main subject. So the boom should enter a set from the opposite side to the key (the two forming an angle of say, 90–120°). The shadow is then thrown to the side of the subject. But it may still be visible in any wide shot taken from the same side of the set as the key light. Wide-angle shots should therefore be taken from an angle well away from the key light; the camera taking them will be close to the boom.

In order to allow for boom shadows to fall on to the floor rather than a wall, it is best to allow at least 5ft between any important action and the backing (in fact, this makes back lighting easier too). In addition, the boom arm should not be taken close to any wall or vertical feature that is likely to appear in vision at that time. Take it too close, and even soft-fill lighting from a wide source forms a shadow.

Shadows from static microphones
Slung microphones can also cause shadows – and especially if they are close to a wall; but such a shadow may be easier to conceal (provided that the surface is not plain and evenly lit) because it does not move. It follows that if a boom shadow is unavoidable, it may similarly be concealed in some feature of the background provided that it does not move while in shot.

Where lighting and sound are at odds, this should be discovered long before a production reaches the studio. The director will arbitrate or may find a solution by changing the action or asking for a change of set layout or design. In practice, boom-shadow problems rarely result from bad planning. They are such a headache that everyone makes an effort to avoid them. But they can arise from a succession of minor changes and compromises which finally snowball into a bigger problem.

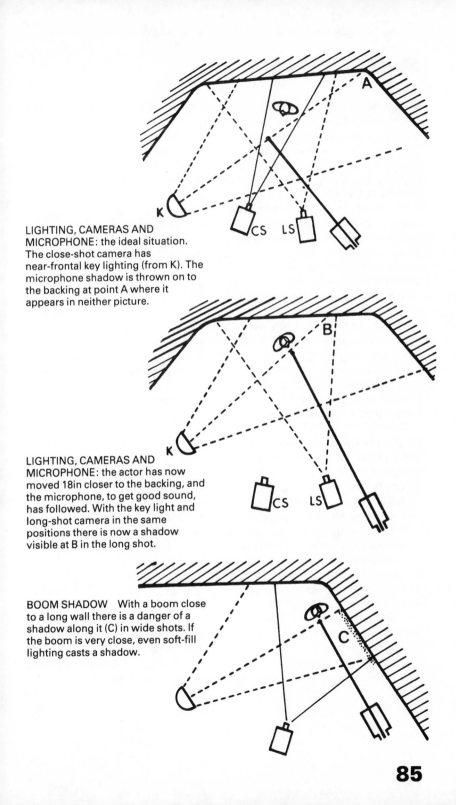

LIGHTING, CAMERAS AND MICROPHONE: the ideal situation. The close-shot camera has near-frontal key lighting (from K). The microphone shadow is thrown on to the backing at point A where it appears in neither picture.

LIGHTING, CAMERAS AND MICROPHONE: the actor has now moved 18in closer to the backing, and the microphone, to get good sound, has followed. With the key light and long-shot camera in the same positions there is now a shadow visible at B in the long shot.

BOOM SHADOW   With a boom close to a long wall there is a danger of a shadow along it (C) in wide shots. If the boom is very close, even soft-fill lighting casts a shadow.

85

Music balance

A 'natural' balance uses the acoustics of the studio, blending these with direct sound by the judicious placing of a single microphone. Additional microphones are added only to remedy faults of internal balance, to cope with difficulties in the acoustics or to bring forward individual instruments, voices or groups (particularly soloists) which are felt to require a higher proportion of direct to indirect sound. Characteristically this type of balance is used for classical music.

The alternative is electronic, or multi-microphone balance. Each instrument or small group has its individual microphone. Many of the instruments are acoustically separated from each other, and the studio is dead, so that little of the reflected sound from one instrument is picked up on the microphones set for the others. The microphone's own frequency response together with equalisation (i.e. filtering) and dynamic compression – all forms of distortion – are used as positive elements of the finished sound. Artificial reverberation is added individually to each channel. This type of balance is generally used for pop music.

The differences between these two approaches are at their greatest when there is a large number of instruments. For smaller groups and single instruments, the techniques overlap: here the characteristics of individual instruments dominate the choice of balance. In the following pages the techniques for music balance are considered instrument by instrument, building up to successively larger combinations.

Checks for music balance
In music balance (of all types) listen for:
1. Wanted direct sound, i.e. sound radiating from part of the instrument where the vibrating element is coupled to the air. It should comprise a full range of tones from bass to treble, including harmonics, plus any transients which characterise and define the instrument.
2. Unwanted direct sounds: e.g. hammer (action) noises on keyboard instruments, pedal operation noises; piano-stool squeaks, floor thumps, page turns, etc.
3. Wanted indirect sound: reverberation.
4. Unwanted indirect sound: severe coloration.
5. Unwanted extraneous noises.

In addition, check that bass, middle and top frequencies are present in the same proportions and are in the same perspectives as each other. This should hold not only for the full, blended sound, but also for each separate component that can be individually distinguished. If so desired, different instruments may be in different perspectives.

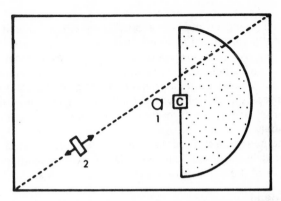

STUDIO LAYOUT FOR BALANCE USING NATURAL ACOUSTICS   A
relatively small amount of the available space is occupied by the musicians
groups round the conductor, C. Position 1 gives a tight, brilliant sound,
discriminating against acoustics that may be over-bright, but may distort the
balance in favour of the nearer instruments and, if a directional microphone
is used, those toward the centre. The microphone will be cardioid or similar.
Position 2, with a bi-directional microphone set back on a diagonal from a
low front corner to a top rear corner gives a well-blended and evenly
balanced sound in which the acoustics play a generous part. Coloration due
to dimensional resonances is minimised by having the microphone off the
centre line.

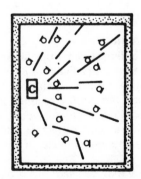

STUDIO LAYOUT FOR MULTI-MICROPHONE BALANCE (on the same scale
as the studio above). This contains the same number of musicians in a much
smaller space. The walls have acoustic treatment to minimise reverberation,
and the sources are separated and sound diffused by the many screens. For a
large group (as here) the screens are arranged so that all players can see the
conductor or leader, C. The actual layout of instruments, screens and
microphones varies from group to group.

87

A music studio may have naturally live acoustics; if it is dead they may be enhanced.

The 'live' music studio

In a music balance using normal live acoustics we are reproducing the characteristics of the studio just as much as those of the players. With a stringed instrument we hear the sound as modified by the sounding board. The studio is a further stage in this same process: it is the sounding board to the whole orchestra and its shape and size give character to the music. But whereas by uniformity of design the characters of all instruments of any particular type conform to a common ideal, those of music studios differ widely.

In the following pages we shall concentrate on the characteristics of the instruments, but it is assumed that the balancer will concern himself equally with those of the studio – about which less can be said.

Too live or too dead

A studio that is too live requires a narrow, close balance, with perhaps a cardioid for a spread source such as an orchestra, or a bi-directional microphone for a concentrated one. (Although a bi-directional microphone is open at the rear, the total live solid angles are less than for the cardioid.) In a studio that is deader than the ideal a wider polar response (up to and including omnidirectional) is employed, or the microphone is placed at a greater distance from the source.

Air humidity strongly affects acoustics. There is usually little attenuation in air below 2000Hz. But at 8000Hz, air at 50 per cent relative humidity has an absorption of 0.028 per foot, resulting in a loss of 30dB at 100ft (or rather less if the air is very damp). In very dry air, however, absorption is greater still and extends down to lower harmonics (400Hz) as well. Performances in the same hall on successive days may sound very different; and close-balanced sound in a dead studio will have different instrumental quality as well as different acoustics.

Ambiophony in the dead studio

A general-purpose television or film studio is the dead studio on the grand scale: its size may be similar to that of a live music studio or concert hall, but if it is to be used for purposes other than music it is much more dead. Typically it has a reverberation time of 0.7 seconds as against, say, 1.3 seconds or more for music. Such a studio would be unpleasant for orchestral musicians to play and and may affect the internal balance of the music, because the players can no longer evaluate their own contribution by ear.

In such a studio ambiophony may be used: this is a technique in which the sound is continuously recorded (on a close balance – with microphones 6–8ft from the source, if possible), delayed, and fed back through fixed loudspeakers on the walls and lighting grid of the studio to simulate reflected sounds. The ambiophony circuit should be kept quite separate from that for the main programme: the microphones it employs will generally be too close for good balance.

CURVED SURFACE   Concave architectural features focus sound selectively emphasising particular components (and often noise). Microphones should not be placed within the radius of a vault such as this.

TELEVISION STUDIO WITH AMBIOPHONY LOUDSPEAKERS
A – scattered around the walls; B – at lighting grid or gantry level. C, The orchestra below. The loudspeakers (0–4) feed delayed sound back into a studio which would normally be too dead for the acoustic balance of music. The balancer treats the result as a normal live music studio, but keeps his microphones well away from individual loudspeakers. This is a typical layout: each actual case may be different.

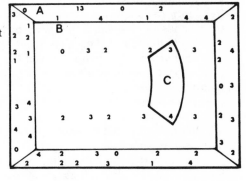

AMBIOPHONY DELAY CIRCUIT   R, Recording head, fed by ambiophony microphones M. S, Frequency response shaping filter. D, Delay heads: typically with (1) 30 and 60 milliseconds, (2) 90 and 150ms. (3) 120ms, (4) 180, 210 and 240ms or (0) zero delay: direct feed to loudspeaker. F, Feedback loop and volume control to give delays longer than 240ms. T, Tension pulley. E, Erase head.

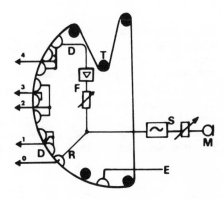

With stringed instruments we relate the microphone to the sounding board, not to the strings themselves.

The string family: the violin

Vibrating strings do not radiate sound directly: they slice through the air without moving it much. They are coupled through a bridge to a wooden panel – a resonator or sounding board – and it is this which radiates the sound.

The rear panel of the irregularly shaped box that constitutes the resonator of instruments of the violin family is shielded and sometimes also damped by contact with the body of the performer; and the sides of this box are fairly stiff. The high-frequency radiation is therefore emitted most strongly nearly at right angles to the front panel, a little off-axis towards the E string. Only the lower frequencies, whose wavelength is comparable to the size of the body of the instrument, are radiated in anything approaching an omnidirectional pattern, and even for these the more directional upper harmonics are important.

Microphone position
The audience does not normally sit in the line of this strongest high-frequency radiation: from the violin, it travels upward and outward over their heads to resonate in the body of the hall itself, and to soften and blend with the sound of its fellows and with other instruments. The sound of a solo violin – including much of the original harshness, squeak and scrape that even a good player emits – is substantially modified before it is heard by an audience.

In positioning a microphone for the violin the sound balancer may prefer the extra brilliance that a high, close balance (not usually less than 3ft) will give him, but he should not think that this is how the violin is supposed to sound.

For a concert violin balance the microphone is set well away from the instrument and usually off its line of maximum high-frequency radiation. Experiments with balance can be made by the usual technique of direct comparison tests, switching between two (or more) similar microphones set in trial positions, the better balance being retained and the other changed for further comparison tests.

Microphone frequency response
The lowest string of a violin is G, of which the fundamental at 196Hz is weak because the body of the instrument is too small to radiate this efficiently (although the overtones are strong). Very high frequencies (say, above 10 000Hz) are also not so important, for the reasons already given. So a microphone with a limited range is satisfactory for strings. Ideally it should have a slightly greater range than the instrument, but if it is very much greater it also picks up unwanted sound either from other instruments or simply as noise. A ribbon microphone is satisfactory for strings.

VIOLIN BALANCE   1, Low
frequency radiation. 2, High
frequencies. 3, Move along
this arc for more or less high
frequencies. 4, Move closer
for clarity; more distant for
greater blending.

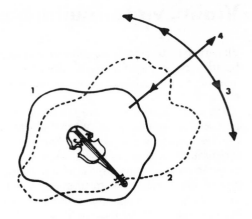

CLOSE BALANCE FOR
THREE DESKS OF VIOLINS
The microphone is above
and a little behind the first
desk. For very clean
separation between violins
and other instruments it may
be necessary to work closer,
with a separate microphone
for each desk, but this
produces a very strong high
frequency response which
must then be corrected.

EXTREMELY CLOSE BALANCE IN POP
MUSIC   The player has encircled the
microphone stand with his bowing
arm. A directional microphone is not
subject to much bass tip-up, as the
lowest frequency present is a weak
196Hz. In fact, a filter at 220Hz may be
used to discriminate against noise and
rumble from other sources.

Violin, viola, 'cello, bass

Violas can be treated in the same way as violins, except that in a very close balance on a directional microphone (e.g. when used as a novelty instrument in popular music) there is some bass tip-up that requires correction.

The larger instruments are progressively more omnidirectional in their patterns of radiation, but still have a forward radiating lobe of upper-middle frequency components. For deeper-toned body from a mass of instruments, a sideways balance is satisfactory; for richness of tone a position closer to the axis is required. For a group in which all of the strings are present the microphone position is largely conditioned by violin balance, although the microphone should be able to 'see' all of the instruments: with the 'cello in the second row of a string quartet the body of the instrument should not be shielded from the microphone. Listen to each instrument in turn as though it were a soloist. Aim to combine clarity and brilliance of individual instruments with a resonant, well-blended sonority for the group as a whole.

In a fairly close balance for a 'cello or bass the microphone is in front of the body of the instrument, perhaps slightly favouring the upper strings. For the 'cello the response of the microphone should be substantially level to below 100Hz at the working distance – or equalised for this – and lower still for the bass. In an orchestral balance, basses may need an additional close microphone, set at about 3ft from the front desk.

The double bass in popular music

When the double bass is used in the rhythm section of a jazz, dance, novelty or pop music group, a very close balance discriminates against other instruments of the group. Be aware that the music stand may act as a reflector. The following positions are all possible:

1. A cardioid, supercardioid or figure-eight microphone set a few inches in front of the bridge and looking down at the strings.
2. A similar microphone near the f-hole by the upper strings.
3. A 'personal' or small studio microphone suspended from the bridge.
4. A 'personal' microphone wrapped in a layer of foam plastic, suspended by its cable *inside* the upper f-hole.

Balance number one, probably the best, picks up the percussive attack quality as the string is plucked, and the ratio of this to the body resonance can easily be controlled. Balance number four, the least likely option, gives maximum separation, but a heavy hanging quality which has to be held to a lower level when it is mised with the other instruments.

One supercardioid microphone which may be used for the first three of these balances has a response which falls away in the bass, so that correction for close working is automatic. Most directional microphones will, however, require low-frequency equalisation at these distances.

STRING QUARTET   Typical balance for the two violins, viola and 'cello on a single microphone which for this frequency range may be a bi-directional ribbon. The rich upper harmonics of the 'cello lack the harshness of those of the violin, so the microphone should be in the 'cello's radiating lobe.

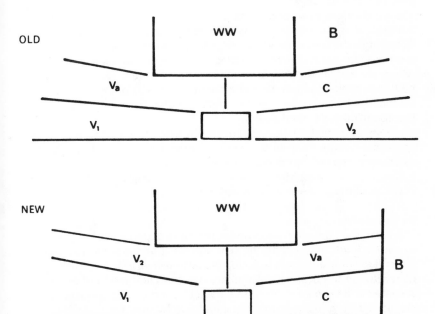

ORCHESTRAL LAYOUT   The more modern form of layout for the strings of an orchestra is based on a revised view of the needs of internal balance which depends on the radiation pattern of the individual instruments both for direct sound and for the way in which the reverberation blends within the hall. The second violins no longer radiate toward the back of the stage. To give the 'cellos and basses more body and presence they have been brought forward. A further possibility is to have the 'cellos in their old position with the violas in front of them, and the basses again to the side.

The concert grand piano

The radiation pattern from the soundboard of a piano allows the balancer a reasonable degree of control over the proportions of treble and bass that he picks up. In particular, the bass is at its strongest if the microphone is placed at a right-angle to the length of the piano; and from this same position the extreme top is clearest if the axis of microphone is directed towards the top end of the soundboard.

From this position we may imagine an arc extending from top to tail of the piano. As the microphone is moved along this arc, the bass is progressively reduced until it reaches a minimum at the tail. For very powerful concert pianos this tail position may be the best, but with a slight disadvantage in that with reduced volume there is also a loss of definition in the bass. For most purposes, a point somewhere in the middle of this arc is likely to give a good balance. This is a good starting point for comparison tests.

Microphone placing

The height of the microphone should allow it to 'see' the strings (or rather, the greater part of the soundboard), which means that the farther away it is, the higher it should be. However, if this is inconvenient, other balances are possible: for example, by reflection from the lid, as in the balance which an audience hears at a concert. For the very lowest tones, the pattern of radiation from the soundboard tends to be omnidirectional; but for the middle and upper register, and the higher harmonics in particular, the lid ensures clarity of sound.

In twentieth century classical music the piano is sometimes used as a percussion instrument. In this case, it is set well back in the orchestra, often with the lid removed. Unless percussion transients need to be enhanced, no microphone is necessary for the piano, even when other sections of the orchestra are picked out.

Action noise

The closer you get to an open piano, the more apparent are the transients associated with strike tone; at their strongest and closest they may be difficult to control without reducing the overall level or risking momentary distortion on the peaks. Action noise – the tiny click and thud as the keys are lifted and fall back – may also be audible, and in balance tests this and the noises from pedal action, etc, should be listened for.

The frequency range of a piano is such that a ribbon microphone can give satisfactory coverage; in addition the bi-directional response gives good fine control of the ratio of direct to indirect sound. Other microphone types with a similar frequency range but different patterns of directional response may also be used.

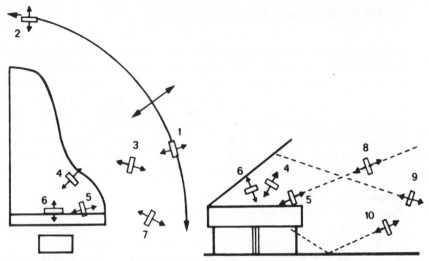

PIANO BALANCE   The best balance for a grand piano is usually to be found somewhere along the arc from the top strings to the tail. A close balance gives greater clarity; a distant balance gives better blending. Of the positions shown: 1, Often gives a good balance. 2, Discriminates against the powerful bass of certain concert pianos. 3, Picks up strong crisp bass (a mix of 2 and 3 can be very effective). 4, Close balance for mixing into multi-microphone dance band balances (with piano lid off and microphone pointing down toward upper strings). 5, Discriminates against piano for pianist/singer. 6, (Angled down towards pianist) as 5. 7, One of a variety of other positions which are also possible. 8, Concert balance 'seeing' the strings. 9, By reflection from lid. 10, By reflection from floor: remember that microphones set for other instruments may inadvertently pick up the piano like this. Percussive transients from the strings are absent.

PIANO AND SOLOIST ON ONE MICROPHONE   Balance for the piano first, and then balance the soloist to the same microphone. The piano lid may have to be set in its lower position.

The piano as rhythm

A rhythm group consists of piano, bass and drums, each being balanced separately. The piano lid is removed, and a directional microphone suspended over the top strings – perhaps as close as six inches. A cardioid or supercardioid will usually be preferred but, surprisingly, even a bi-directional ribbon microphone can be used, because the bass tip-up at this distance actually helps to rebalance the sound. The position chosen will depend on the melodic content of the music: one criterion is that the notes played should all sound in the same perspective. Listen for distortion (particularly on percussive transients) due to overloading.

Close balance

For a percussive effect the microphone could be slung fairly close to the hammers. It has been known for a baffle – perhaps a piece of cardboard or hardboard – to be fastened to the upper side of a ribbon microphone in order to further emphasise and harden the higher frequencies, and especially the transients.

If action noise is obtrusive, use the microphone's directional response to discriminate against it. A second microphone may be added over the bass strings; this can have its own separate equalisation and echo. Another possibility (in which action noise can be completely avoided) is to place one or two microphones a few inches above holes in the frame of the piano. The quality of sound from individual holes can be sampled by ear.

Upright pianos

For upright pianos, listen for each of the wanted and unwanted sounds listed on page 86, again checking that the balance remains the same over the whole scale (except for the cases – with lighter styles of music – when you may prefer to have 'not too much left hand').

Lift the lid and try the microphone somewhere on a line diagonally up from the pianist's right shoulder – a good balance can generally be found in this position. But remembering that it is the soundboard that is radiating, an alternative is to move the piano well away from any wall and stand the microphone on a stool or box at the back (for a close balance), or diagonally up from the soundboard (for a more distant one). At the back, the sound will have body but will lack percussive transients.

One of these positions may also be suitable for balancing a 'jangle-box', a piano which has leaves of metal between hammer and strings (to give a tinny strike action) and in which the two or three strings for each note of the middle and top are slightly out of tune with each other.

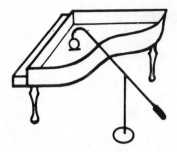

CLOSE PIANO BALANCE   Directional microphone suspended over the top strings – but not too close to the action.

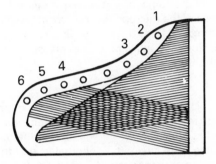

PIANO HOLES   Some pianos have holes like this in the iron frame, and their focusing effect can be used for a pop music balance. A microphone 2 or 3 inches over hole 2 probably gives the best overall balance, but those on either side may also be used, as also can mixtures of the sound at hole 1 with 4, 5 or 6.

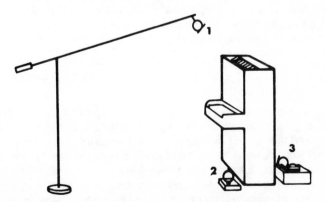

UPRIGHT PIANO   Position 1, microphone above and behind pianist's right shoulder. Position 2, to right of pianist below keyboard. Position 3, at rear of soundboard.

More piano balances

The piano is intrinsically very much more powerful than the human voice, so it is part of an accompanist's job to see that a singer (or any other soloist) is not swamped by the piano sound. But the accompanist's judgement is valid only for the sound in the concert hall; when it is recorded for replay through a single loudspeaker system, his level must be lower still, because our ears are no longer able to discriminate spatially between the two components. (In stereo, the spread of the piano and reverberation helps.)

Balance on a single microphone can often handle this adequately – simply by adjusting the relative distance between the second performer and the microphone. For a concert given before an audience the soloist can stand in the curve of the piano, facing outward: this has the piano and soloist on the same axis of the microphone, but in this position the soloist cannot see and react to cues from the pianist. In a studio performance it would be better for the two players to be on opposite sides of a bi-directional microphone.

Two microphones

A balance using two microphones gives the balancer separate control of the two sources: their relative levels can be adjusted precisely without repeated requests to the musicians or movements of the microphone. In carrying out the balance tests, first get a good balance of the soloist and then fade in the second microphone until it just brings the piano in to the same perspective. This minimum setting will often be quite loud enough.

Where a performer is singing or talking at the piano he needs a close microphone for his voice. A cardioid or any other directional microphone with its dead side toward the piano often picks up quite enough volume from the instrument for an accompaniment. This requires a microphone with smooth polar and frequency characteristics. Without that, the sound quality will be better if a second microphone can be used for the piano. Two pianos may be balanced on one or two microphones in a variety of ways, depending on the studio layout and the preferences of the players (see opposite).

Backing track

A technique that is sometimes used in pop-music recording is to make a backing track. The accompaniment is pre-recorded and the singer (or other performer) listens on headphones to replay, adding his contribution as the combined sound is re-recorded. This technique can be used to obtain special effects: for example, to combine very loud and very soft sounds or where the replay of the pre-recording is to be at a different speed. Note that artificial reverberation should be added to any speeded-up sound if it is to match the original.

PIANO AND SOLOIST ON TWO MICROPHONES – I Microphone 1 covers both instruments and provides the studio reverberation. Microphone 2 adds presence to the soloist's sound: here a bi-directional microphone has its dead side to the piano.

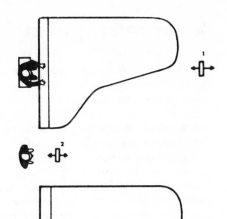

PIANO AND SOLOIST – II A second version of two-microphone layout. Here the pianist and soloist can see each other well.

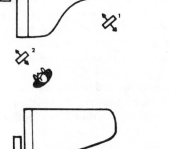

TWO PIANOS – I Here a single microphone is used, giving an adequate if not perfect balance. The players can see each other's faces.

TWO PIANOS – II Balance is on a single microphone, 1, or individual microphones, 2 and 3. The players can see each other's hands.

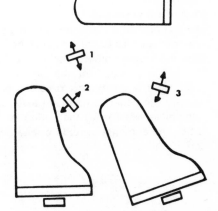

A miscellany of strings

The classical guitar radiates sound in much the same way as the middle members of the violin family, so a microphone forward of and above the player achieves a reasonable balance. Noises arising from the fingering – rapid slides along the strings and so on – are characteristic of the instrument and need not be eliminated entirely, but in a more distant balance they are less pronounced.

The folk guitar, banjo and ukelele are all used as accompaniment to the player's own singing. In these cases the first step is to balance the voice: often, with little adjustment, this will give adequate coverage of the instrument as well. If an additional microphone is required it may be placed either close to the body of the instrument (in which case the same stand may be used), or farther back to give a somewhat more distant coverage to include both singer and instrument.

Other instruments of the same general class are the viola, lute etc; the harp; and also the balalaika and the mandolin. The mandolin often benefits from added perspective; in a group of instruments its penetrating tone and continuous sound overlay the reverberation and make it sound more forward than it actually is.

Other keyboard instruments

For a close balance on keyboard instruments such as the harpsichord, clavicord or virginals first identify the sound board (or box) and balance the microphone to that, discriminating, where necessary, against mechanical action noise. In a wider balance that includes other instruments the harpsichord may be too loud and penetrating. Its position should be chosen so that if necessary a directional microphone can be angled to hold the other instruments, while discriminating a little against the harpsichord. The celeste presents the opposite problem: it is rather quiet, and unless the music is written to favour this, an additional close microphone may be required to reinforce its sound.

Exotic instruments

The balancer may be faced by many other instruments. These may be regional (such as the zither); ancient; exotic or oriental; or newly invented. A good starting point for any unknown instrument is to put a microphone above the position of the audience in the acoustics for which the instrument is designed (some are designed for the open air) and then employ the usual comparison tests. For a closer balance start from first principles, working to the sound radiator, avoiding harsh axial sounds which the audience would not hear, and discriminating against intrusive action noise.

HARP   The soundbox is cradled by the player. Part of the action – the pedals for selecting the key of the instrument – are at the foot, F. A balance from microphone position 1 gives adequate coverage from a distance, but for closer work the less obvious position 2, above and slightly behind the player's head, may be adopted. In a very close balance (e.g. when the harp is used as a novelty instrument in a light-music group) the microphone may be directed toward the soundbox.

CELESTE   As with the upright piano, a microphone near the soundboard behind it avoids mechanical noise which can be loud compared with the music in this relatively quiet instrument. A close balance is often needed to get adequate separation from other, louder instruments of the orchestra. An alternative position for the microphone is in front of the instrument, below the keyboard, on the treble side of the player's feet at the pedals – and keeping well away from them.

Woodwind

For woodwind, with fingered or keyed holes, the bell radiates relatively little of the sound – and very much less than in the brass. Woodwind sound is emitted largely through the first few open holes. It follows that unless we need to go very close we have considerable freedom of position in balancing woodwind: a microphone placed somewhere in front of and above the players should prove satisfactory.

Transients

Distance softens the transients that are produced by the initial excitation of the sound by edge tone or reed. These transients are characteristic of instrumental quality, and their exaggeration in a close balance is no great disadvantage – with the exception, perhaps, of the flute and piccolo which can sound like wind instruments with a vengeance at close quarters. But in balancing these, the player's head can be used to shield the microphone from the edge tone.

Saxophone characteristics

Where woodwind is employed in popular music – particularly in a dance or show band – it often takes the form of a saxophone section. (Note that the defining characteristic of 'woodwind' is the method of playing and not, despite the name, the material of the instrument.) Up to five saxophones may be employed, some of the players doubling on clarinet, flute or even violin.

On a stage these would be laid out in a line or arc with a microphone to each desk of players, with an additional microphone for any soloist. In the studio this might become a tight semi-circle round a downward-angled cardioid. An alternative is to group them three and two on either side of a central bi-directional microphone. A ribbon has a suitable frequency response, but its live angles project horizontally and so are in danger of picking up open studio reverberation and the sound of other instruments – particularly from the brass. A pair of downward-angled supercardioid microphones avoids this. Also the broader body of the saxophone permits more radiation from the bell, which the microphones pick up directly, thereby allowing the lowest possible control volume setting and so the greatest separation from other instruments. In this last balance an additional microphone is again required for any soloist.

To exaggerate the characteristic quality of the woodwind in a popular music balance, and to allow its 'presence' to be heard through other instruments, midlift may be employed. A peak of 5–8dB, centred on 2500–3000Hz, is introduced electronically.

WOODWIND   Typical layout in an orchestra. F, Flute. P, Piccolo. O, Oboe. C, Clarinet. B, Bassoon. These are generally raised on rostra in the centre of the orchestra, behind the strings. If the main orchestral microphone is set for a close balance, it may tend to favour the strings at the expense of woodwind, and an additional microphone may be required. This can be a bi-directional microphone pointing down at about 45°, with its dead side toward the instruments in front of the woodwind section. Alternatively, set two supercardioids closer in – about 3ft up from the flutes and oboes.

FLUTE   In this position the microphone can still 'see' the pipe and fingerholes, so the balance is good. Placing the microphone behind the player's head shields it from the windy edge tone produced as the player blows across the mouth hole of the instrument.

WOODWIND IN DANCE BAND (SAXOPHONES)   In a studio layout up to five saxophones may be grouped for a close balance around a central microphone position. *Left:* a bi-directional microphone set at the height of the bells of the instruments. *Right:* C, Two cottage-loaf microphones set above bell level and angled 45° downwards. This reduces pick-up from outside the woodwind area. S, Solo microphone for player doubling on a second instrument, so that this can be controlled separately.

Brass

In a brass instrument the bell emits the main stream of high harmonics, but the audience will not usually hear these directly unless the music specially requires that trumpets or trombones are directed toward them. Most of the time these instruments are angled down toward the music stands or floor. The tuba is directed upward to the roof, and the horn backward away from the audience. In general, a good balance hears the sound much as the audience does – with a microphone forward of the brass and not too high. In the orchestra, with the brass set back and raised behind the woodwind, this is accomplished satisfactorily on the main microphone.

The horns benefit from acoustic reinforcement by reflecting screens set behind them. For a special if rather uncharacteristic effect, and almost complete separation on a close balance, the microphone may be set behind a horn player and in line with the bell.

Allowing for increases of volume

With trumpets, and even more with trombones, the volume can be high. If the output from a close microphone is fed directly to preamplifiers of fixed gain these may overload and cause distortion. 'Pads', fixed or preset attenuators, should be inserted before the amplifier. The balancer should not rely on rehearsal levels of 'big band' brass players he does not know, but should allow for as much as a 10dB rise in volume on the 'take'. Failure to do this could leave him without an adequate working range on the faders, and face him with new and unrehearsed problems of separation.

The range of volumes that includes both open and muted brass can be wide. A close balance on trumpets and trombones is easier to control if the players are prepared to co-operate by leaning in to the microphone in the muted passages, and sitting back for louder passages.

For the trombones the appropriate height for a microphone conflicts with that of the music stands: the leader of the section can be consulted on whether the microphone should be above or below the stands.

In popular music, brass is another group that may benefit from midlift somewhere in the region of 6000–9000Hz. Some older microphones already have a peak in this range.

Brass band

A brass or military band may be laid out either in an arc or along three sides of a hollow square, with the cornets and trombones facing each other at the sides and the deeper-toned instruments at the back. Balance can be achieved on a single microphone placed a little way back from the band and off the centre line of the studio.

104

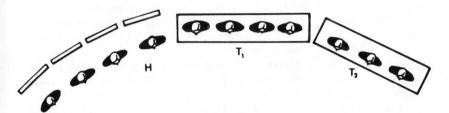

ORCHESTRAL LAYOUT OF BRASS   H, The horns need a reflecting surface behind them. T_1, The trumpets need to be raised high enough to see over the heads of the woodwind. T_2, The trombones, at about the same level.

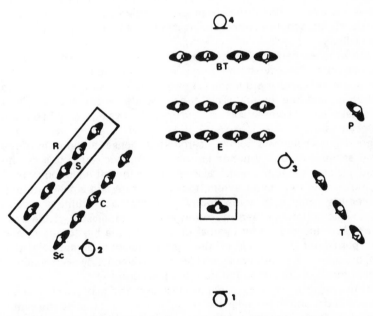

BRASS BAND   Possible layout and microphones. 1, Main microphone: cardioid. C, Cornets. S, Sopranos (ripieni). The rear rank is set on a rostrum R as they tend to play downward into the backs of the line in front, thereby losing echoing figures. Sc, Solo cornet playing to microphone 2. T, Trombones. P, Percussion. E, Euphoniums; soloist comes forward to play to microphone 3, which points downward to separate him from other instruments of similar sound quality. Euphoniums and bass tubas, BT, radiate directly upward: on the main microphone the quality sounds woolly unless there is a good reflecting surface above them. If not, use microphone 4 above the bass tubas. This can be a ribbon; the others, condensers.

*An orchestral section that really needs high-performance microphones –
and the most demanding part of a pop group.*

Percussion, drums

It is this department more than any other that demands audio equipment
with a full frequency range and a really good transient response. The bass
drum produces high acoustic power at low frequencies, while the triangle,
cymbals, gong and snare drum all radiate strongly in the extreme high
frequencies, i.e. in the octave above 10 000Hz (although 15 000Hz is a
reasonable limit for broadcasting.

As the microphone moves closer to the percussion, it becomes necessary to check the balance of each element in turn as it is used, adding
further microphones as required. A vibraphone may require a close
microphone beneath the bars, angled towards the open taps of the tubular
resonators.

At the back of an orchestra, percussion is usually set up on rostra. These
need to be solidly built, or reinforced, in order to avoid coloration.

Balance for pop percussion

In popular music, percussion shows itself mainly in the form of the drum
kit: this includes snare drum, bass drum (operated by a foot-pedal),
cymbals, hi-hat (a foot-operated double cymbal), and tom-toms (medium-
sized drums).

In the simplest balance for these a single microphone points down at
the snare drum. But any attempt to move in for a close balance from the
same angle increases the dominance of the nearer instruments over the
bass drum and the hi-hat. (The hi-hat radiates most strongly in a
horizontal plane, in comparison with the up-and-down figure-eight pattern
of the cymbals.)

A good close balance employs a cardioid condenser at the front of the
kit and at the level of the hi-hat. For an even tighter sound this can be
moved in closer still to the hi-hat, also favouring the left-hand top cymbal
or cymbals and, to a lesser extent, the snare drum and small tom-tom,
together inevitably with some sound from the bass drum; a second
cardioid microphone is then moved in over the other side of the bass
drum to find the remaining cymbal or cymbals, the big tom-tom, and
again, perhaps, the snare drum. Changes of position, angle, polar diagram, or relative level give considerable control over nearly every element
– except for the clarity of sound from the bass drum.

For the bass drum set a microphone close to the edge of the skin where
there is a wider and more balanced range of overtones than at the centre
of the skin. This is one instrument for which an old heavy moving-coil
microphone with a limited top-response is as good as a modern high-
quality microphone – which might in any case be overloaded. For a dry
thudding quality, remove the front skin (tightening up the lugs to prevent
rattling) and put a blanket inside.

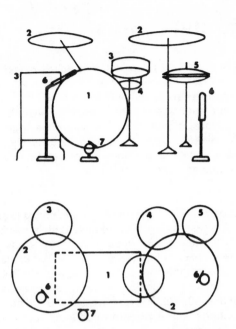

TYPICAL DRUM KIT for dance band or pop group. 1, Bass drum. 2, Cymbals. 3, Tom-toms. 4, Snare drum. 5, Hi-hat. 6, Microphones with good high-frequency response used in two-microphone balance. 7, Additional microphone (moving coil) for more percussive bass drum effect (used in some pop music). If further microphones are used, the sound from the snare drum may benefit from midlift at 2800Hz, plus additional bass and extreme top.

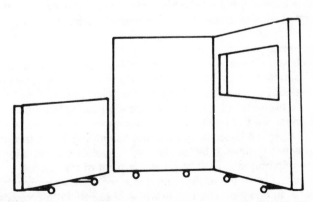

STUDIO SCREENS   Absorbers about 4in thick. A dead music studio has a supply of full and low screens that are easily moved on rollers to provide acoustic separation without loss of visual contact. A drum kit will often have full screens to the sides and low screens at the front. A vocalist's screen may have a window of stiff transparent plastics.

For clarity of diction, singers are generally balanced close.

Singers: solo and chorus

What counts as a close balance for a singer depends on the type of music. For a pop song it may be 2in; for an operatic aria it may be 4ft. In either case this is a very much closer perspective than you would hear in the concert hall or theatre. In a public performance with no microphone, the demands of internal balance require that the singer be loud or the accompaniment soft. But where a close microphone is used, there is full freedom to create brilliant orchestral backings which can be interwoven and contrasted with a subtle and varied vocal line.

Frequency response
A microphone with a supercardioid (or figure-eight) response is satisfactory, with equalisation for bass tip-up if necessary. With a microphone which has bass roll-off to compensate for close-working, bass may have to be added if the singer moves back. A frequency range limited to that of the voice helps to avoid spill on to this microphone at the upper and lower ends of the orchestral frequency range. In popular music, presence may be added somewhere in the 1500–3000Hz range. Sibilance can be reduced by selection of a microphone with a smooth middle-top response, with high-frequency roll-off if necessary; and perhaps by singing at an angle across the microphone.

Artificial reverberation may be added to a close balance, but clarity of the words (if clarity is required) limits this to less than that of the instrumental backing. For a backing chorus several singers group tightly round a directional microphone. For a larger chorus, the group gathers round a cardioid microphone, which can still be positioned to give good separation from the louder instruments. The singers at the sides of the group can come closer to the microphone than those at the centre, their line following that of the curve of the microphone's polar diagram.

Balancing a chorus
In balancing a larger chorus for serious music – opera, oratorio or other choral music – aim first for clarity of diction (except in those cases where a more distant balance is specially called for). The microphone must be far enough back and high enough to get a well-blended sound but the limit is generally the lower limit of intelligibility. One or several cardioid, supercardioid or figure-eight microphones are used, depending on the layout of the singers. These may have to be too high up for soloists to work to; for them, separate microphones may be placed in front of the chorus.

There is no particular virtue in separation into registers – except where forces are weaker in one than another, in which case separation may help the balancer to compensate for lack of internal balance. In general, allow the group to follow its established procedure.

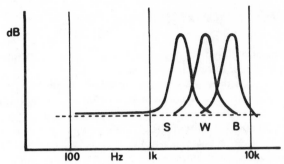

PRESENCE PEAKS   These are midlift – added electronically at different frequencies. Singer S, woodwind W and brass B each have their separate range within which their character is selectively emphasised, so that one does not conflict with or obscure another.

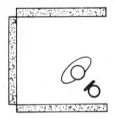

SCREENED AREA FOR VOCALIST   When the voice level is much lower than the volume of surrounding instruments the microphone is screened from them. The live side of the microphone is towards the screens, the dead side towards the open studio.

CHORUS   A high, directional microphone favouring the rear singers gives an even balance. A figure-eight or cottage-loaf response (perhaps using several microphones in a line) gives coverage to the chorus while picking up little sound from the area in front of it (which may contain an orchestra).

Orchestra – and organ

The full orchestra and the pipe organ both have a large number of individual sound sources. But each can be treated as a single composite instrument, and as such they have similar problems. Both have a very wide range of frequency and volume and so demand a high-quality microphone.

Both are spread in space, so the microphone – or microphones – must be placed so that the internal balance is not unduly distorted. But this need not be taken to extremes: no member of an audience could be equidistant from all of the individual sources.

Both have as their sounding board the hall itself, so that the main problem of microphone placing is to balance direct with reflected sound. And in both cases the character of the composite instrument can be severely affected by the presence of an audience. But they, too, can be treated as part of the instrument: their number and placing can sometimes be arranged with the balance in mind. A recording that is to be issued on tape or disc would be marred by coughing, but for a broadcast performance a live audience is good.

The acoustics
Historically, both orchestra and organ have been fashioned by the very nature of the surroundings in which they are likely to be heard: the balancer may feel that he is fighting against the acoustics, but he must never forget that the music was written not only for this particular combination of sound sources (with all its possible minor variations in performance) but also for acoustics which are unlikely to be too different from those he is faced with.

But orchestra and organ differ in that the combination of organ and the dominant characteristics of its acoustic environment is fixed, whereas the orchestra can be varied in layout and size, and even physically moved – in the final resort to another hall.

A second possible difference is that the acoustics surrounding the organ may be more resonant, and perhaps more subject to echoes. In these circumstances the acoustic effects of the absence of an audience and even of the humidity or dryness of the air is more noticeable than in the concert hall.

But in most cases it should be possible to find a satisfactory microphone position. In his search, the balancer needs not only technical skill but also an attitude of mind: he is testing for ways, not of beating the acoustics, but of making them work for him as effectively as possible.

Other than for television, many orchestral balances are in stereo, either with a coincident pair or with spaced microphones. For a full account of this, see *The Technique of the Sound Studio* (Focal Press, 1979). Additional 'spotting' microphones are usually mono (even in a stereo balance), as described here.

110

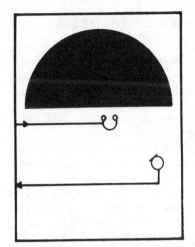
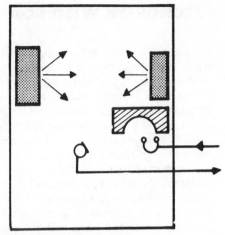

HEADPHONES Neither the conductor *left* nor organist *right* is ideally placed. The conductor is too close, and the organist may be in an awkward corner or out of line of sight of some of the sections of pipes. When they are producing sound specifically for the microphone, headphones *may* possibly help them to modify their performance for the better.

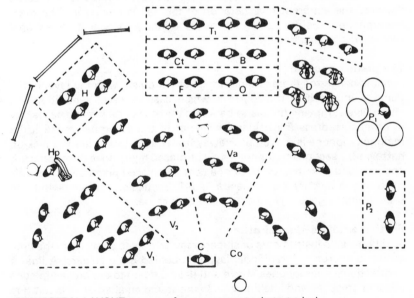

ORCHESTRAL LAYOUT may vary for many reasons, but a typical arrangement may be something like this. V_1, First violins. V_2, second violins. Va, Violas. Co, 'Cellos. D. Basses. H, Horns. T_1, Trumpets. T_2, Trombones. F, Piccolo and flutes. O, Oboes. Ct, Clarinets. B, Bassoons. P_1, Tympani. P_2, Other percussion. Hp, Harp. The main microphone may be augmented by others for particular sections or instruments (shown here for harp and woodwind), but only if these are necessary. *First* see what can be done by layout, *then* correct for any imbalance that remains.

Orchestra with soloists or chorus

In a concert layout a single soloist is usually placed at the conductor's left hand. A single-microphone balance for the orchestra may give satisfactory results for the combined sound: again, this is how the audience hears it. The music is written with the expectation that the soloist will be only a little forward of the orchestra, but the microphone can slightly accentuate this if placed a little off-centre or angled to favour this position.

Many listeners prefer to hear a balance that brings the soloist out a little more than this; and, especially in mono, it is desirable to separate him from other sounds of similar quality. An additional closer microphone favouring the soloist can be faded in until the soloist is just appreciably in closer perspective, but beware of other instruments which may also be in its field. In an alternative studio layout, the soloist is separated by being brought out to the front and side of the orchestra. Where directional microphones are used, each may then be placed with its dead side to the other source: this gives the greatest separate control of the component sounds. The value of a closer balance on the soloist comes mainly from the slightly increased high-frequency content.

When a second or third microphone is added at much the same level as the first, the apparent reverberation of the hall is increased. The main microphone therefore has to be moved a little closer, or its directional characteristics changed.

The piano concerto

In piano concertos there should not be any problem with the level of the piano – unless it is too loud, in which case physical separation (or, alternatively, moving the piano back into the orchestra) may be necessary in order to redress the balance in favour of the accompaniment. If the main microphone is moved back away from the orchestra (and its pick-up narrowed) a second microphone may be placed high above the conductor, and shielded from the full blast of a centrally placed piano by its lid. This microphone favours the orchestra, and in particular the woodwind, the section that is generally most in need of such benefit.

Further soloists – and chorus

Additional solo instruments or singers may be treated similarly to the first, but for a chorus the traditional layout is behind the orchestra. A line of bi-directional microphones above and dead-side-on to the orchestra adds missing presence and intelligibility. In the studio an alternative is to bring the chorus to the front and to one side of the orchestra and to balance it separately.

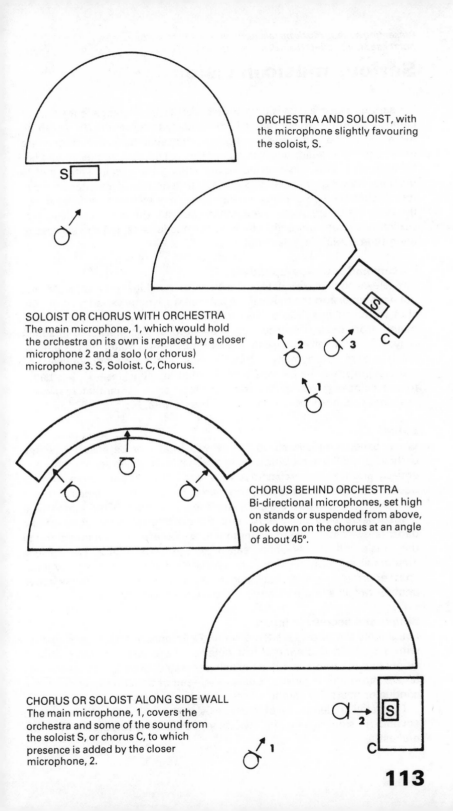

ORCHESTRA AND SOLOIST, with the microphone slightly favouring the soloist, S.

SOLOIST OR CHORUS WITH ORCHESTRA
The main microphone, 1, which would hold the orchestra on its own is replaced by a closer microphone 2 and a solo (or chorus) microphone 3. S, Soloist. C, Chorus.

CHORUS BEHIND ORCHESTRA
Bi-directional microphones, set high on stands or suspended from above, look down on the chorus at an angle of about 45°.

CHORUS OR SOLOIST ALONG SIDE WALL
The main microphone, 1, covers the orchestra and some of the sound from the soloist S, or chorus C, to which presence is added by the closer microphone, 2.

113

*Presenting serious music on television may entail compromises –
but these should affect the balance as little as possible.*

Serious music in vision

For serious music a television picture makes little difference to the type of balance sought – except that it is more likely to be mono. But it can be more difficult to achieve a good balance because television studios are generally rather dead; because microphones and their stands should not be obtrusive; and because the layout should be visually satisfying. In practice, however, normal concert layout generally gives adequate pictures; while the use of concert halls rather than television studios solves the problem of adequate reverberation as well. Directors and designers sometimes adopt more visually adventurous layouts, but the success of their attempts is often debatable.

Picture and sound perspectives

The television camera can take close-ups of individual orchestral players, but it would distort the music if the balance were to be varied from shot to shot to match this. The balance required for television may, however, benefit from more presence than is usual for music presented in sound only: this will justify the close-ups and may sound better on the inferior loudspeakers in so many television sets. If for dramatic rather than musical purposes the balance is modified to match a close-up, it is better to add a close-up microphone to a normal balance, rather than to present the whole sound from the position of the camera.

Ballet

When ballet is presented on television it may not be easy to get the orchestra into the same studio. In this case it is as well to separate the two entirely and take the orchestra to a concert hall or sound studio where a good balance can be obtained, and then to feed this by loudspeaker to the dancers. Indeed, even if the orchestra is in another part of the same studio loudspeakers may still help them. Sound effects in the form of the dancers' steps or rustle of costume that are occasionally audible through the music will add life and realism to the performance. For these a directional microphone should be placed on the dead side of any loudspeaker array. (A set of small loudspeakers arranged in a line one above another radiates in a horizontal plane, with a dead zone above them.)

Singer and accompaniment

For singers the principal alternative to a microphone in vision is a boom with a cardioid microphone just outside of the edge of frame. Where singer and accompaniment are widely separated, there is a slight time lag unless the singer is given a loudspeaker feed of the accompaniment (the conductor hears the combined sound on headphones).

The relative deadness of most television studios can be turned to advantage: singers generally require less artificial reverberation than the orchestra.

114

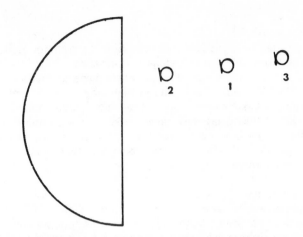

ORCHESTRA IN VISION   Where a standard balance for sound only is
achieved on microphone 1, a combination of the other two microphones may
be more suitable for television. Microphone 2 gives clarity but insufficient
reverberation; this is added by microphone 3.

BALLET: PLAYBACK AND EFFECTS   The dancer needs to hear the music
from reasonably close at hand, so a loudspeaker may be necessary. Here a
series of loudspeakers set in a vertical line radiate high frequencies in a
horizontal field; a microphone above them receives only low-pitched sound.
This can be filtered out, as the main component of the required effects is at
higher frequencies. Close microphones are ruled out by the very wide shots
that are often necessary in showing ballets.

Opera

Opera may be recorded or broadcast in concert form, i.e. without movement; or it may be taken direct from a stage performance.

In a studio performance the simplest layout has the chorus forward of and to one side of the orchestra. There is a row of microphones in front of them, to which the soloists also work. (Curiously, such a layout works in stereo, too; the action being laterally reversed so that the corners of orchestra and the chorus that are closest are on the same side of the stereo sound stage: indeed, many of the techniques that have been described for mono continue to be useful in stereo.)

Problems of stage opera

For stage opera good balance is not easy. In mono it is customary to use a high microphone to get a good balance on the orchestra together with a row of cardioids in the footlights (dead side towards the orchestra) to pick up the singers. An alternative that gives greater separation is 'microphone mice', microphones set in foamed rubber pads on the front of the stage itself. If the singers are well back there is no problem of balance between several voices (though they may appear too distant). But if a singer comes down stage very close to a microphone his footsteps can become gargantuan: he is therefore balanced on the next microphone along in the direction that he is facing.

If two singers are working together and one is well down stage of the other, the best microphone to balance them may be on the opposite side of the stage: this will be roughly equidistant from both, and should still see the face of the downstage singer even if he turns partly away from the audience.

One of the main problems here is the excessive distance of voices in most positions of the stage. It is therefore best to avoid having a long row of open microphones: they make coverage easy, but only at the expense of adding even more reverberation. Fade down those not in use.

One or two slung cardioid microphones, high above the orchestra and directed to favour the singers, may give a better balance. If there is a paying audience these will generally have to clear the sight lines to the highest part of the auditorium.

Sound and the stage play

The techniques for mono opera from the stage may also be suitable for plays *televised* from the stage. The convention of stage voice projection is thoroughly unsatisfactory for television, but audiences will accept it and especially in comedies where the theatre audience is more obviously reacting. But it is not suitable for plays recorded or broadcast in sound only, as theatre acoustics are far too lively. A recording may be made as a matter of historical record, but there is no method of coverage that is both satisfactory and totally unobtrusive.

OPERA IN THE SOUND STUDIO
1, A cardioid microphone picks up the orchestra on a fairly close balance: with so many other microphones open, there is no lack of reverberation. 2, Microphone for announcer, narrator or sound effects. 3–6, Microphones serving soloists and chorus. The conductor, C, is set back from his usual position.

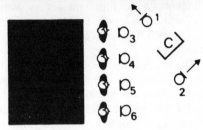

MICROPHONE POSITIONS FOR THE STAGE (e.g. for opera)   1 and 2, A pair of cardioid microphones in the footlights. Though very close to the orchestra their response discriminates against it. 3, A single cardioid microphone placed high but well forward in the auditorium.

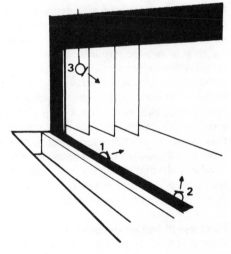

MONOPHONIC OPERA BALANCE FROM STAGE PERFORMANCES   Singers A and B, well down stage, are balanced on microphone 3 or 4, but not on 2 which is closer. For stereo such a balance would also be better, but entails reconstruction of their positions on the sound-stage.

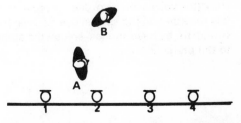

117

In a production employing vast musical resources,
the main problems are sound communication and sound spill.

Televised opera

Opera in the television studio may be presented with actors miming to music prerecorded or performed in another studio; with the singers in a television studio and orchestra in a sound studio; or with singers and orchestra both in the same hall or studio.

Miming is probably the least satisfactory method, although it has been employed successfully. Its main advantage is that it permits the camera to concentrate on plot and action without being limited by the artificial convention of opera. The technique is similar to that already described for ballet, with the same care to pick up significant sound effects and spoken passages.

The second technique generally carries more conviction and spontaneity. The singers are covered by booms, etc, as for a play, but there are new problems of communication between the separate studios for orchestral sound and television picture. The interchange of picture – including that of the conductor, who needs to be visible to the singers or repetiteur – is technically straightforward, but the provision of sound links is complicated by the need to avoid spill.

Sound links between singers and orchestra

Orchestral sound is relayed to the action studio floor by means of a vertically mounted line-source loudspeaker on the front of the boom dolly. As this is moved around to cover different action areas, the spatial relationship between loudspeaker and microphone is maintained, and good separation can be achieved. The voices of the singers can be relayed to the conductor by a partly shielded loudspeaker over his head or by a line loudspeaker on the back edge of his music stand. Line loudspeakers used for communication should not be allowed to radiate bass at frequencies below those corresponding to their length: this ensures that their directional qualities do not deteriorate and cause spill at low frequencies.

Where the orchestral and acting areas are at opposite ends of the same hall, similar methods may have to be employed to overcome sound delays due to the length of the studio.

Post-synching an orchestra

Where scenes must be prerecorded or filmed before the orchestra is available, a separate guide track of the beat or piano accompaniment is recorded. On film this can go on the combined optical or magnetic sound track (the voices are recorded on separate magnetic sound). On videotape it is recorded on the special cue track. In the studio the cues are fed to the conductor by headphones, so that the orchestra can then be synchronised to the prerecording.

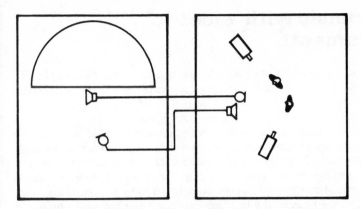

COMMUNICATION between sound and television studios that are used together presents obvious possibilities for the degradation of sound quality. Precautions taken to avoid significant spill include the use of line loudspeakers which are placed so that they do not radiate strongly in the direction of the microphone.

OPERA OR BALLET ON TELEVISION with the conductor and orchestra in a separate studio from the action. 1, Line loudspeaker along the top of the music stand: three small (3in) eliptical loudspeakers produce little bass and are adequately directional at high frequencies. An alternative is 2, a loudspeaker in a hood. The 4ft square baffle limits high frequency spill but may also get in the way of lights. 3, Monitors for conductor to see action. 4, Camera to relay beat to repetiteur in television studio.

Pop music with 'electric' instruments

Most modern pop music is created by the treatment and mixing of close balanced sound. A common component of this is the 'electric' instrument with a loudspeaker. This sometimes also appears as a composite instrument, as in the case of the electric/acoustic guitar.

In the electric guitar each metal string vibrates in the field of a coil. The vibration induces an alternating signal which is amplified within the body of the instrument. Sometimes there are several rows of signal generators. To obtain a signal from a conventional guitar or from a violin or bass, a contact microphone (transducer) is fixed, perhaps by double-sided adhesive tape, to the body of the instrument. Placed just below the bridge, it responds to and sums the vibrations from all of the strings. The output of the transducer or guitar pre-amplifier is passed through an external amplifier to a loudspeaker.

Loudspeaker characteristics

This loudspeaker is usually capable of handling high power (in order to provide strong bass) but need not necessarily be of high quality: the character of the instrument as a whole includes that of the loudspeaker. For the most flexible balance, the feed from the guitar is split, with one line going directly to the mixer and the other to a loudspeaker from which the sound is picked up by microphone and also fed to the mixer. The two components have separate equalisation and reverberation.

The loudspeaker of an 'electric' instrument may be balanced like any other instrument: for fullest high-frequency response the microphone should be on the axis of the high-frequency cone. A supercardioid microphone may be used for a close balance, or a figure-eight (ribbon) where the loudspeaker is to be balanced with the acoustic output of the same instrument on opposite sides of a single microphone.

The pop group of the 'sixties consisted of one drum kit, three electric guitars plus the voice of one of the guitarists. A studio layout that achieves reasonable separation has the loudspeakers in line on a common axis each with a close directional or supercardioid microphone. The guitarists are on one side of this line and the drummer is on the other, on the dead side of the loudspeaker microphones and with as much screening as proves necessary.

Equalisation for vocalist

The microphone for the vocal line may be equalised for very close working, with presence added by midlift at 2000–3000Hz and perhaps compression as well if the voice is supposed to blend into the backing. If the vocal is compressed this means that the backing can be brought up higher behind it. The backing may in turn be limited to hold back any excessive peaks: this will therefore result in a high common level for all of the sound.

120

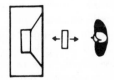

ACOUSTIC/ELECTRIC GUITAR balanced on a single bi-directional microphone. The amplifier-loudspeaker unit is set on a box opposite the player, with the microphone between loudspeaker and instrument. The player himself can now adjust the overall amplifier level, comparing it with the direct, acoustic sound: he has control of the loudspeaker volume from moment to moment by means of a foot control. If separate microphones are used, they can be treated separately, e.g. by 200Hz bass roll-off on the acoustic guitar channel.

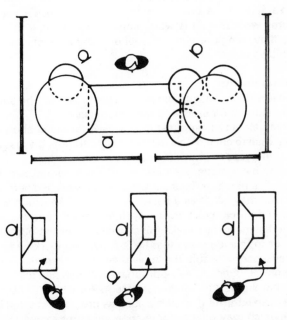

POP GROUP with three electric guitars, drums and vocalist. Maximum separation has been achieved by screening the drummer heavily: there are two low screens in front of him, high screens at the sides, and a further screen bridged across the top. The two main microphones for the drums are at the rear, and the bass drum microphone is separated from the skin only by a layer of foam plastic. For studio safety the power supplies to electric instruments should be fed through isolating transformers.

121

Rhythm group and small band

We have already seen how piano, bass and drums are balanced individually (pages 96, 92 and 106). We may now put them together, and add further instruments: a guitar (acoustic/electric), brass, woodwind (primarily saxophones), a vocalist, and so on. Instruments used for particular effects might include celeste, violin, horn, harp, electric piano or anything else that interests the arranger. A low screen helps to separate the sound of bass and drums. Further screens are required, e.g. between piano and brass, and around particularly quiet instruments or singers. For good sound separation a dead studio is chosen whenever possible. In the dead acoustics of a big recording studio, the various components are often placed in screened booths around the walls. Lead and bass guitar loudspeakers may be directed into half-screened bays and taken by cardioid microphones. An acoustic guitar may require a separate room. Considerable use is made of tracking (sequential recording), particularly for vocals.

When bright surroundings are all that is available some spill may be inevitable. In this case we may have to modify the ideal fully separated balance. For example, if the brass is going to spill on to the vocalist's microphone come what may, it is worth considering whether to make it good-quality spill by using the same microphone: the vocalist sings at 3in and the brass blows at 6ft.

Factors affecting layout

The layout of instruments, microphones and screens within the studio is affected by the nature and purpose of the group itself:

1. The group may already have a set layout used for stage shows. The balancer may wish to change this to improve separation. Discuss this in advance with the group's musical director.
2. The group may have made previous recordings or broadcasts in the same studios. The balancer finds out what layout was used before, and if it worked well may use it as a basis for his own. The group will not expect to be in different positions every time they come to the studio. However, the balancer may make minor changes for the purposes of experiment, or to include ideas that he is convinced work better than those already used. Consult the individual instrumentalist, section leader or musical director, as appropriate. Major changes may be dictated by the use of additional instruments (of which the balancer should have been informed), or particular new problems created by the music itself (which may only become apparent in rehearsal).
3. The group may be created for this session only, and the balancer – having found out the composition from producer or musical director – can start from scratch. But he must look not only to the needs of balance; he must also understand the problems of the musicians.

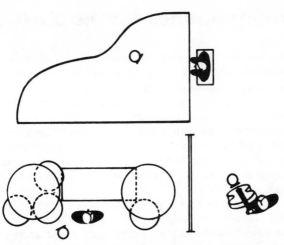

RHYTHM GROUP   This is the basic layout to which other instruments may be added. One low screen is required to separate bass and drums. Various positions are possible for microphones for piano (see page 97), bass (page 92) and drums (page 107), with additional microphones for the drums as necessary.

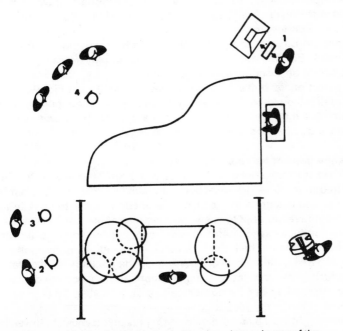

SMALL BAND   To the rhythm group (with microphones in any of the positions already described) are added: 1, Guitar (acoustic/electric). 2, Trumpet (with a full screen to separate this from the drums). 3, Trombone. 4, Woodwind. (This is one of many possible instrumental groups.)

Many microphones for the big band

Where the studio layout is decided or influenced by the balancer he is guided by principles which are designed to help the musicians and must achieve his balance within their limits.

1. The drummer, bass player and pianist form a basic group (the rhythm group) which expects to be together.
2. The bass player may need to see the pianist's left hand, to which his music directly relates – and especially when they improvise.
3. When the bass player has fast rhythmic figures he needs to hear the drums, particularly the snare drum and hi-hat; these can be heard well on the drummer's side.
4. In a piano quartet the fourth player can face the other three from beyond the piano: everybody can see everybody else. But a guitar player may have to work closely with the bass: placed at the top end of the piano keyboard, he can see behind the pianist to the bass player.
5. Any other two players who are likely to be working together on a melodic line or other figures need to be close together.
6. Where players double, e.g. the pianist against playing a celeste, the two positions are together.
7. All players, including a saxophone (and woodwind) group which may be split with players facing each other, must be able to see the musical director or conductor.
8. The musical director may also be a performer playing a particular instrument. But if it has been arranged that all players can see each other, condition 7 should still be satisfied.

The application of these conditions may appear to conflict with the original idea of building up a layout according to good principles of separation. But the careful use of directional microphones and screens should make good sound possible.

The balance cannot be heard in the studio

The only satisfactory studio is one that has been specially built, with acoustic treatment to reduce reflections to much less than would often be used in a studio designed for speech. It is uncomfortable to work in for musicians who are not used to it, and no clear idea of the result is heard by a leader or conductor who does not wear headphones. Indeed, even with headphones he may not hear the final result if the individual components are being fed to a multitrack recorder, because the final mix is made later, following further experiment with equalisation, artificial echo and mixing (although, depending on the number of tracks available, there may be partial mixing during the recording).

A session might start by recording rhythm plus guide vocal, followed by acoustic guitar and piano, and next the lead vocals and backing. Then come solo guitar and riffs; an extra instrumental line may be added by an earlier performer.

124

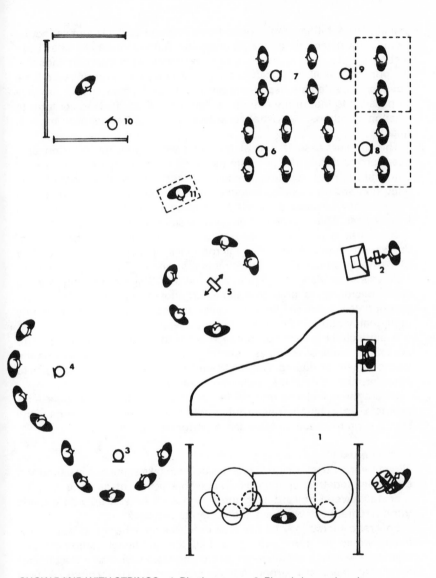

SHOW BAND WITH STRINGS   1, Rhythm group. 2, Electric/acoustic guitar.
3, Trumpets. 4, Trombones. 5, Woodwind. 6, First violins. 7, Second violins.
8, Violas. 9, 'Cellos. 10, Vocalist. 11, Musical director. Variations must be
made for acoustic difficulties: in one layout it was found necessary to bring
the bass forward to the tail of the piano and separate him by acoustic screens
from brass, drums, and woodwind (the trumpets having been moved farther
away). A wide range of different microphones is used: see the notes on
individual instruments for details.

Popular music in vision

Visual considerations may require that the sound studio layout cannot be used for musicians appearing on television. But for a group that has been recorded, the quality of the sound by which they are known demands that the principal results of that layout still be achieved. And groups who have not been recorded are dependent on sound quality, that must be as close as possible to that of the recording studio if their music is to conform to existing standards. So visual considerations must not override musical quality.

The simplest case is that of the band which appears in vision only occasionally during the programme. Here, the techniques of the sound studio are carried over almost unchanged: the fairly dead acoustic of most television and film studios favours the multimicrophone technique. Low screens may be used to divide the sections, and the conductor wears headphones. This is accepted as a visual convention.

Where the players are more strongly featured, the main change from the sound studio is to avoid microphones or stands that are unduly obtrusive, or which obstruct the line of the camera. But sound still comes first, and again visual conventions permit this. Singers generally work to stand microphones or with hand microphones; but where this is inappropriate a boom can be used, as close as possible, and perhaps with a high-quality gun microphone to improve separation.

A particularly rewarding approach is for the director and designer to accept the needs of sound as a starting point, and to attempt to make the layout visually intriguing. For this to work well, the sound man, too, has to enter into the spirit of the co-operative venture.

Where the visual and musical demands are totally incompatible – for example, an exacting dance routine where the dancer also sings – the sound is prerecorded and the dance performed to playback.

Filmed music

Film differs from television in that it is usual to take shots one at a time and edit them together afterwards. The whole action of a scene may be repeated several times, taking the sound each time, if only as a guide track. Separate cutaways are shot, often to playback.

Several cameras may be used, as in television, when repetition of the music would be more expensive than the additional equipment and operators, where a unique event cannot be reshot, or where the director feels that there are strong artistic benefits to be obtained. A final possibility is that there will be more music played (e.g. at a public event) than is finally required on film, so that cutaways can be shot by the sound cameraman during this same performance. The beat can be marked out in the guide track, and appropriate actions matched to that of the master sound.

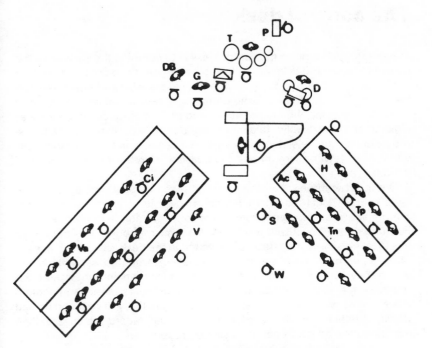

SHOW BAND appearing in vision and giving support to singer or other featured acts. Typical layout: V, Violins. Va, Violas. Ci, 'Cello. DB, Double Bass. G, Acoustic guitar. T, Tympani. P, Percussion. D, Drum kit. H, Horns. Tp, Trumpets. Tn, Trombones. Ac, Accordion. S, Saxophones, doubling woodwind. W, Overall wind microphone. All microphones may be condenser cardioids except for moving coil on bass drum.

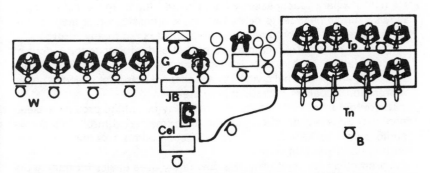

ANOTHER POSSIBLE LAYOUT FOR DANCE BAND   Again the rhythm is in the centre to give the strongest beat in such an excessively spread layout. W, Individual microphones for Saxophone (woodwind) section. JB, Jangle box piano. Cel., Celeste. G, Guitarist and amplifier. Tp, Trumpets. Tn, Trombones. B, Brass microphone.

The control desk

The primary function of the control desk is to combine the various sound sources at appropriate levels. At its most complex, the input to a television desk may include up to 60 microphones, several disc and tape reproducers, reverberation plates or echo rooms, and, say, 16 outside source lines (from telecine and videotape, other studios, remote locations, overseas circuits and the public telephone system). There may, typically, be provision for the selection of, say, 40 sources at one time. A radio desk would have fewer microphones, but possibly more sound reproducers of various types.

The input volume may range over as much as 90dB, grouped as low level sources (such as microphones) or high level (such as lines from remote sources). All signals are raised to a common high level at about 50–70dB above the output level from most microphones. This minimises electrical noise and the effects of induction in the desk itself and in lines to other equipment. Each source is channelled through a separate fader, which may be in the form of a rotary knob (convenient when there are relatively few to control) or quadrant or slide faders (easier to handle if there are many). Where several microphones are likely to be faded up and down together they are preset on individual faders which are then combined in and controlled by a single group fader.

Additional features of the desk

The desk may also have provision for modifying the frequency response on a number of channels ('equalisation' and compression) and for feeding some proportion of each signal to the echo system. Filters to reduce full frequency sound to telephone quality for dramatic purposes may be switched in. In television these may be interlinked with picture cut buttons. A press-button 'pre-hear' facility allows circuits to be sampled on a small auxiliary loudspeaker without fading them up. A 'solo' facility allows a single channel to be monitored, suppressing other feeds to the loudspeaker but without affecting the mix that is fed to any recorder.

A foldback switch (which may be pre-set and left) feeds the output of selected faders to a studio loudspeaker for cue (or possibly mood) purposes; and additional switch in each channel taps it for a feed to public address loudspeakers in the audience area.

The output from some of the channels may be split to provide a 'clean feed' that does not pass through the main (master) control. This allows studio output to be fed to a remote location without sending back the contribution from that source.

A prompt-cut (or cough key) is a now rarely-used device for muting the studio microphone output from a position in the studio itself.

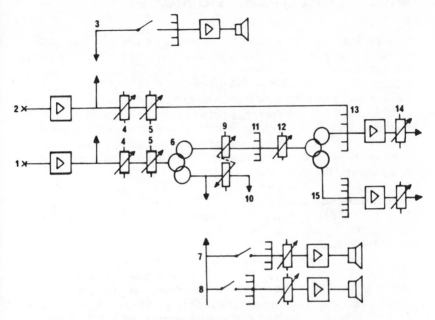

MICROPHONE CHANNEL (radio or television studio)   1, Normal microphone input. 2, Independent microphone (signal to be mixed in after group faders). 3, 'Prehear' side chain. 4, Microphone balance attenuator (preset). 5, Channel fader. 6, Hybrid transformer, splitting the signal. 7, Foldback side-chain (to studio floor loudspeakers). 8, Public addres side-chain (to audience loudspeakers). 9, Echo mixture switch. 10, Echo feed. 11, Star mixer combining a group of sources. 12, Group fader. 13, Star mixer combining groups and independent sources. 14, Main gain control. 15, Clean feed chain (supplying feed of all groups except independents). In addition there may be a response selection amplifier (for equalisation) and a compressor/limiter.

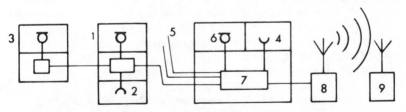

THE BROADCASTING CHAIN   The studio output may go to a recording room, to another studio or direct to a programme service continuity suite. Between studio, continuity and transmitter there may be other stages: switching centres, boosting amplifiers, etc. 1, Studio suite. 2, Tape recording room. 3, Remote studio. 4, Tape reproducing room. 5, Other programme feeds. 6, Continuity announcer. 7, Continuity mixer. 8, Transmitter. 9, Receiver.

Modifying the audio signal

The facilities provided on and after the control desk may be used either to combine sounds in a form that is as natural as possible (that is, by using the studio acoustics to enrich and blend the sound), or to creatively modify the sound and blend it in the control desk itself. Some balancers have strong preferences for one technique or the other, but their partisanship reflects their specialised skills or musical interests. Both techniques are valid, as are combinations of the two. There are many ways in which the facilities of the control desk and intermediate recordings are used actively in balance. Response selection (equalisation), compression and echo are discussed later.

Building sound stage by stage

Tracking, building up a complete sound by additive recording, is taken a stage further in post-balancing, or 'reduction'. Broad tape is used to record many tracks (typically 16 or 24) in synchronisation so that the recording session itself can be devoted to establishing the broad musical qualities and the basic contribution of each individual instrument. The precise details of treatment and relative levels are worked out later. This is a powerful technique which has been widely used in the creative balancing of pop music. It allows the balancer to concentrate on one thing at a time and to experiment. He begins by checking the treatment given to individual tracks, starting with the rhythm, and can then gradually assemble a composite sound. If the result is unsatisfying he can try again: the original tracks are not affected.

Noise reduction system

In a recording studio producing commercial records or tapes, noise arises in any intermediate tape recording. A noise reduction system is therefore used in conjunction with many of the techniques described above: indeed, in an analogue signal system it makes them possible. Essentially, in such a system, the frequency range is split into several parts. The frequency bands with least energy (and in which noise would be most noticeable) are automatically raised in level before recording, and reduced by a corresponding amount afterwards as a coded signal controls the reconstitution of the original sound.

Noise can also be avoided by converting the signal to digital form. This is particularly useful for master recordings.

Editing may also be employed and involves physically cutting tape, or copying extracts from the original to a second tape.

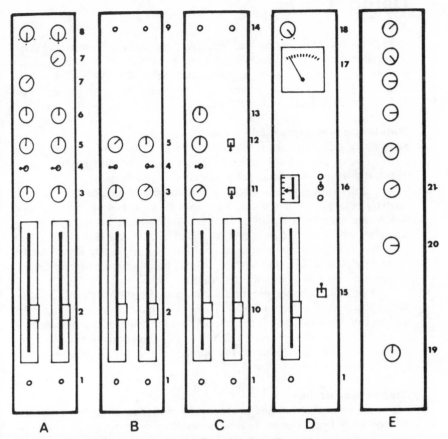

MODULAR UNITS FOR COMPLEX CONTROL DESK As each module carries two channels the basic units can also be modified for stereo sound by clipping the faders together in pairs. A, Individual source module (two channels). 1, Prehear buttons. 2, Faders. 3, Public address volume control. 4, Selector switch: p.a. before or after fader. 5, Feedback volume control. 6, Echo feed control. 7, Preset attenuator for coarse setting of volume from source. 8, Group selector switch. B, Group control (accepts all individual faders switched to it, plus group echo output). 9, Emergency change-over push button for group amplifier (using public address feed amplifier as spare). C, Master control module. 10, *Left:* clean feed fader. *Right:* main fader. 11, Public address 'cut' switch. 12, Foldback feed 'cut' switch. 13, Prehear overall volume control. 14, Main amplifier emergency change-over push button. D, Echo control module. 15, Echo 'cut' switch. 16, Echo plate controls (to change reverberation time) with scale. 17, Peak programme meter, indicating volume of signal sent to echo room or plate. 18, Switch selecting output from one of two groups as echo feed. E, Audience Mixer. 19, Volume control for loudspeaker in audience area. 20, Group fader for audience microphones. 21, Six faders for audience microphones (this is a low-level mixer in which the relative levels should be preset). Each channel for a microphone or other input will also have provision for 'equalisation', modification of the frequency response.

Using filters

Filters may be active or passive. A passive filter can only reduce the level of the signal passing through it, doing so selectively by frequency. With the simplest circuit the filter passes substantially all of the signal on one side of a chosen frequency and cuts progressively more and more of it on the other side of this turn-over point.

Simulating telephone quality, etc.

Simple switchable passive filters can be used to simulate telephone quality sound, public address, intercom systems, and loudspeaker output within the action of a dramatic presentation. The original signal comes from a studio microphone which is separated and shielded from other open microphones to avoid spill. For complete separation it can be placed in a closed cubicle, with the performer wearing headphones. In a television play where he has to appear in vision in alternate shots, this is not possible: in this case the studio layout must be designed to minimise the possibility of spill. Directional microphones help to only a limited extent, as much of the spill is by reflected sound.

The frequency range of most public telephone systems is some 3000Hz, with a lower limit of about 300Hz. There is no need to copy this exactly: the degree of cut-off used is a matter for the judgment of the balancer, as is the relative level of the 'far' voice in a telephone conversation (meters provide no useful indication of the appropriate level when the bass has been removed from a sound).

Response selection

For more complex changes in frequency response an active filter employing amplifiers is used. In inexpensive domestic equipment this can very simply be incorporated into amplifier circuits which are needed anyway, so simple bass and treble controls (tone controls) are widely available. When they are used to select a response that is satisfying to the listener, he exercises the same judgment that the balancer employs more expertly in the process that is commonly called equalisation – but which might more properly be described as response selection.

Nominally, 'equalisation' as applied to microphones means restoring a level frequency response to a microphone that has inherently or by its position an uneven response: it may be used to match microphones with different responses. In practice it may often also mean taking a second step in the creative distortion of sound for a particular effect (the choice of microphone and its position being the first). It is used in this way in pop music – and the result is a matter for the judgment of the balancer.

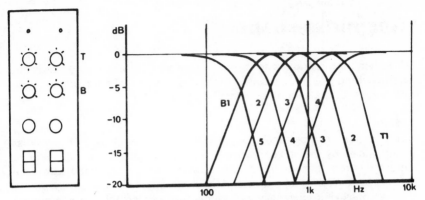

EFFECTS FILTER   *Left:* A module containing two units (as in equipment shown on page 131). *Right:* The simple passive filter has four degrees of bass cut-off (B1–4) and five degrees of top cut (T1–5). In a television control desk these may be used in conjunction with a camera channel switching unit for reversing telephone effects.

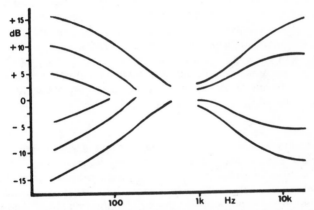

BASS AND TREBLE CONTROLS   In contrast to the effects unit above, these require an active filter (i.e. containing an amplifier) and can be more complex. These response curves are samples only: particular cases may differ widely from this.

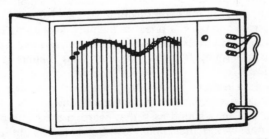

SHAPING FILTER   In this design the slide faders may be at, say, third or half octave intervals and the band centred on each successive slide is attenuated or amplified according to its volume setting.

*More complex filters are used
to change the central part of the frequency response curve.*

Response shaping

A rather more complex form of filter selectively amplifies a particular narrow range of frequencies. Where this is done without reducing the remainder of the sound, the effect is called midlift.

Midlift

This is the electrical analogue of the selective emphasis of particular frequencies by the vocal cavities or the dimensions of the resonator of many instruments. In these, the formant that is produced gives the instrument much of its character. Midlift can therefore be seen as a device for emphasising character and, indeed, if used out of any appropriate context can do so to the point of caricature.

The most common use for midlift is in popular music. Here the emphasis of individual character helps instruments that are battling for recognition in a loud and densely structured sound. As we have seen, singers, woodwind and brass may be given peaks of midlift at different frequencies and so may be combined in a single piece of music. These peaks may also be described as 'presence', as they appear to bring forward a particular source without affecting its overall volume. A comparable effect is a mid-range dip: call it 'absence'.

Digitally-programmed equalisation can have almost continuously variable control of frequency or volume and can also vary the width of the curve (known as the Q).

Shaping filter

If we take response selection one stage further up the scale of complexity we reach the shaping filter. In this the signal is split into bands which may nominally be an octave wide (or for more precise work even narrower). In practice, these will not be discrete bands but overlapping peaks, so arranged that for a 'level response' the sum at any frequency is equal to that of the original signal.

Selection of the desired effect is made by setting the slides for the various component bands at different levels to form what appears to be a graph of the new response on the face of the control unit. The actual response that is achieved may differ from this to an extent that depends on individual design and the sharpness of the fluctuations set. But, in any case, the relationship between the curve that is set and that achieved is not particularly important; the only thing that really matters is what the final result sounds like.

This type of equalisation is particularly useful in matching sound in dubbing film. Voices recorded in different conditions cannot, in general, be matched so that they sound as if they were in the same place. But deficiencies in quality and intelligibility can be made less obtrusive, and unnecessary minor variations can be evened out.

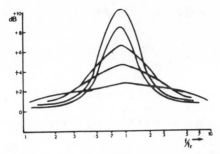

MIDLIFT   Response curves for a midlift tone control unit. In a simple unit centre frequencies $f_r = 2,3,5,8$kHz (approx.) might be selected; also five degrees of lift, as shown. A more complex unit would have much closer, stepped settings, or a slide control for frequency selection.

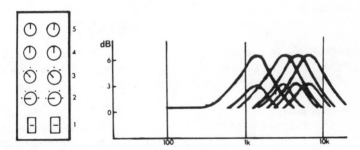

RESPONSE SELECTION AMPLIFIER (EQUALISATION)   A very simple channel desk-mounted module comparable with the equipment shown on page 131. 1, Bypass switch. 2, Mid-lift selection (1.4, 2.8, 4.0, 5.6kHz). 3, Presence control (3 and 6dB lift). 4, Bass lift and cut control. 5, Treble lift and cut control.

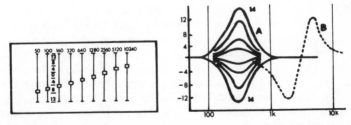

AUDIO RESPONSE CONTROL   Graphic filter for a more versatile control of response than can be provided by bass, midlift, and treble controls only. *Left:* desk control panel. *Right:* A, Typical response curves for a single selector. B, Response obtained by setting adjacent selectors at −14dB and +14dB. The example shown is an octave filter; graphic filters with much narrower intervals are professionally used.

135

Compressors and limiters

Compression (reduction in dynamic range) may be used to maintain a high overall level in order to ensure the strongest possible recorded or broadcast signal. In popular music it may be applied to single parts of a backing sound which would otherwise have to be held low in order to avoid sudden peaks blotting out the melody; or it might be used on the melody itself, which might otherwise have to be raised overall, and as a result become too forward.

How a compressor works
Below a predetermined level (the threshold or onset point) the volume of a signal is unchanged. Above this point the additional volume is reduced in a given proportion, e.g. 2:1, 3:1, or 5:1. For example, if the threshold were set at 8dB below the level of 100 per cent modulation and 2:1 compression selected, it would mean that signals which previously overmodulated 8dB were now just reaching full modulation. Similarly if 5:1 had been chosen, signals which previously would have overpeaked by 32dB will now only just reach 100 per cent.

What this means in practice is that the overall level may be raised by 8dB or 32dB, so that relatively quiet signals are now making a much bigger contribution than would otherwise be possible.

How a limiter works
Suppose now that the compression ratio is made large, say 20:1, and the threshold level raised to something very close to 100 per cent modulation. Working like this the compressor now acts as what is called a limiter. A limiter can be used to hold unexpected individual high peaks that would overload subsequent equipment or it can be used to lift the signals from virtually any lower level to the usable working range of the equipment – though this may cause problems with high level signals that may be excessively compressed, and with background noise that will be lifted as well.

It will be seen that the effect of a 2:1 compressor with a threshold at 8dB below 100 per cent modulation is to compress only the top 16dB of signals while leaving those at a lower level to drop away at a 1:1 ratio.

The decay time in a typical simple limiter may be set at 0.1, 0.2, 0.4, 0.8, 1.6 or 3.2 seconds. The fastest of these can hold brief transient peaks without affecting background but with a rapid series of peaks could produce an unpleasant vibrato-like effect. For most purposes a setting of about half a second is satisfactory, but the duration chosen is not critical at low compression.

Compressors and limiters should not be used for classical music. Inded in any circumstances where manual control is feasible it is almost invariably to be preferred. The use of compression in pop music is a special case.

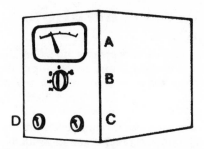

COMPRESSOR/LIMITER   A, Attenuation meter (scale: 0–24dB). B, Threshold setting, −24 to +16dB relative to nominal 0dB setting. C, Compression control, 1:1 (i.e. no compression), 2:1, 3:1, 5:1, and Lim (this is 20:1, i.e. acting as limiter). D, Decay time control: 0.1–3.2 seconds. These are typical ranges for a simple device.

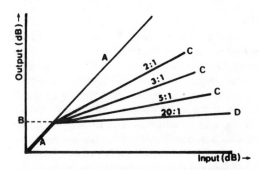

COMPRESSION RATIOS   A, Linear operation, in which input and output levels correspond. Above a given threshold level B, various compression ratios, C, reduce the output level. In the extreme case, D, the compressor acts as a limiter holding the maximum output volume close to the threshold level.

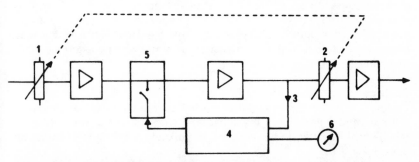

SIMPLE COMPRESSOR/LIMITER   1 and 2, threshold level control (ganged). From the main programme chain a sample signal, 3, is fed to a side chain, 4, that has compression ratio and recovery time controls, then to a control element, 5, operating on the main programme chain. The degree of attenuation is shown on a meter, 6.

137

Echo: artificial reverberation

'Echo' is an odd name for a device which serves to extend reverberation without, it is hoped, introducing any actual echoes – but the name has stuck. It is used on occasions when more reverberation is wanted than the built-in acoustics of a studio can supply, or (for music) when a multi-microphone close balance has been adopted.

Most control desks have facilities for taking a feed from each source to the echo facility. Since each can be set for a separate amount of echo, this allows greater control than would be possible with natural acoustics.

A variety of methods are available: echo chambers, reverberation plates, systems of springs and digital techniques. The first three produce a random decay of sound by a system of multiple reflections; and the main technical design problem in each case is the removal or avoidance of distinctive colorations.

The echo chamber

An echo chamber may be a room with 'bright' reflecting walls, perhaps with 'junk' littered about at random, in order to break up the reflections and mop up excess mid-range reverberation. A humid atmosphere gives a strong high-frequency response; a dry one absorbs top. An echo chamber linked to the outside atmosphere varies with the weather.

The shape and size of the room controls its quality of reverberation. A rectangular shape should be avoided, as its dimensional resonances will produce distinctive colorations, further emphasised by the many reflections required within the given time in a volume that is often rather small (typically 4000 cubic ft or less). A sloping ceiling helps.

Another disadvantage of the echo chamber is that, once laid out, its reverberation time is fixed. Two seconds may be suitable for music but less satisfactory for drama: a little echo goes a long way on monophonic speech. A further problem, which may add to the cost, is that echo chambers need to be isolated from structure-borne noise.

Echo chambers have one major advantage over the convenient and predictable reverberation plates and springs: the decay is natural because the reverberation is in three dimensions. A plate, with two, sounds more mechanical.

Springs

'Springs' would appear to start off with a severe dimensional handicap, but have progressed from the cheap and cheerful (but characteristically mechanical) to offer, in some cases, remarkably lifelike effects for given uses.

The acoustic input is applied torsionally at one end of the helix, and sensed in the same manner at the other. The metal is differentially etched and notched along its length, so that its transmission characteristics change. The mismatches at all the discontinuities create a vast multiplicity of paths.

138

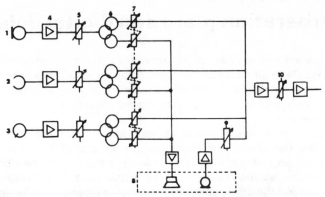

SPLIT FEED SYSTEM FOR ECHO   1, Microphone. 2, Tape reproducer. 3, Record player. 4, Pre-amplifiers. 5, Source faders. 6, Hybrid transformer giving independent feeds. 7, Echo mixture switches (ganged). In a typical system this has nine positions: in the central position both the direct and echo feeds are at full volume. By turning it to the left the echo feed is reduced step by step to zero; turning to the right gradually reduces the direct sound. 8, Echo chamber. 9, Echo fader. 10, Main control.

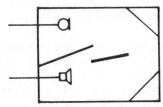

ECHO CHAMBER   A U-shaped room is often used to increase the distance the sounds must travel from loudspeaker to microphone. A sloping ceiling and non-parallel walls prevent the formation of dimensional resonances.

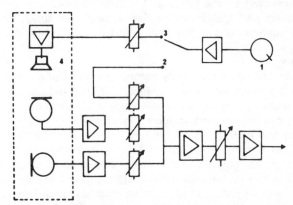

ACOUSTIC FOLDBACK   Tape or disc reproducers, 1, which are normally fed direct to the mixer via switch contact, 2, may sometimes be switched via 3 to provide an acoustic feed to the studio loudspeaker, 4. This may be used where a sound recorded with little or no reverberation will benefit by the addition of some studio reverberation.

139

Reverberation plate and digital delay

The reverberation plate takes up less room than an echo chamber and is less susceptible to structure-borne noise. It can, however, pick up sound from the air, so to avoid howl round it should not be housed in the same room as a loud monitoring loudspeaker. It is not affected by other plates in the same room.

The plate is rather like the sheet of metal once used in the theatre to create a thunder effect, but instead of being shaken by hand and delivering its energy directly to the air it has two transducers. One vibrates the plate, rather as the coil vibrates the paper cone in a moving-coil loudspeaker; and the other, acting as a contact microphone, picks up the signal, reflected many times from the edges.

The medium is a tinned steel sheet suspended in tension from a tubular steel frame at the four corners. To reduce the metallic quality of the resonance to generally acceptable proportions a minimum area of 2 square metres is combined with a maximum thickness of half a millimetre: in steel plate this has good transverse vibrational properties. Unlike the echo chamber, its natural resonances do not thin out to sharp peaks at the lower frequencies, but are spread fairly evenly throughout the audio range.

For a given nominal reverberation time the actual duration is longest at mid-frequencies, with a slight reduction in the bass and rather more in the top (15dB at 10 000Hz), thereby simulating the high-frequency absorption of the air of a room of moderate size. To achieve this, some damping has to be applied: without it, the response of the plate would rise to high values in the extreme bass.

Damping is also used to vary reverberation times between 0.3 and 5.3 seconds, depending on the spacing (3–120mm) between the echo plate and a damping pad that normally has a motorised drive, controlled remotely by push-buttons on the desk, where a dial indicates the setting.

Digital echo

Reverberation can be mimicked by computer control of a digital signal, but needed the development of relatively powerful techniques of manipulation and storage (by random access memory). Given these, sophisticated effects can be obtained, including reverberation times that depend on frequency and unnaturally long times for special effects. Continuous sampling gives a 'spin' effect, which by slightly enhancing part of the frequency range can sound like the rapidly degenerating feedback effect after which it is named. Variable delay is also achieved, and by recombining this with the original signal, 'phasing' and 'flanging' effects can be obtained, a hollow phase-cancellation quality that can be made to sweep through the audio-frequency range. Used as special effects in pop music, this has previously been achieved by manual drag on one of a pair of recordings.

REVERBERATION PLATE   1, Metal sheet. 2, Tubular steel frame. 3, Damping plate, pivoted at 4. (Here the spacing is shown as being set by the hand-wheel, but many damping plates are motor driven from a point on the sound control desk). Two transducers are used: D, Drive unit. M, Contact microphone.

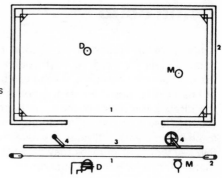

ECHO WITH COATED WHEEL OR TAPE LOOP   1, Input to recording head. 2, Output. 3, Erase circuit. 4, Output mixer. 5, A feed from the last head may be used to continue the sound at progressively lower levels. The quality is characteristically different from echo chamber or plate.

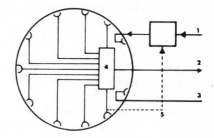

APPARENT REVERBERATION FROM TAPE LOOP   A hard mechanical quality is generally obtained. At lower speeds this opens out first to a flutter effect, then to a series of discrete echoes. Each of these may be useful as a special effect.

PHASING   The change in amplitude (vertical scale) with frequency. The effect of flanging – here with a delay of 0.5 milliseconds – is evenly distributed throughout the frequency range, but is not related to the musical scale, which is logarithmic.

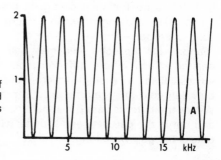

141

Using echo

Artificial reverberation is an integral part of any balance where close microphones are used. As each source has a separately controlled echo feed, differential amounts of echo can be added, depending on the clarity of line or resonance required. Soloists can be brought forward by lifting the direct component of the sound and reducing echo input. The voice requires less echo (either in volume or reverberation time) if the words are to be heard clearly – though in modern music the marginal audibility of a few key words or phrases may be all that is required to establish subject matter or mood. Here the application of echo is a matter for the balancer. In his interpretation of the composer's or arranger's intention there is scope for skill and versatility.

Simulating concert hall conditions

When serious music has been recorded in an acoustic which is too dead (as in most television or film studios), echo may be added in an attempt to improve matters. Because of its size, in a concert hall there is a tiny delay between direct sound and the first reflection arriving at the microphone. What happens in this interval often helps to define the character of particular instruments.

Passing the feed to the echo device through the recording and replay heads of a tape machine reintroduces the delay that would otherwise be missing, and so may add to the feeling of spaciousness. A fast tape speed and close spacing between the heads is needed to simulate realistic reverberation. A delay of a tenth of a second corresponds to a path difference of 100ft, as in a big hall. Note that we are already in the realms of true echo as a delay in excess of a twentieth of a second can begin to separate original sound and reverberation to produce discrete echoes on staccato sounds.

The long path of true concert hall reverberation results (in dry air) in the attenuation of high frequencies. To simulate this the echo output may be fed through a tone control to reduce top. Reverberation times of different durations may be used for singers and orchestra.

Echo on speech

In using echo for speech (e.g. in plays) the main problem is one of perspective: moving closer to a microphone produces more feed to the echo chamber – the very reverse of a realistic effect. A separate microphone for the echo feed may help; so will careful rehearsal of the desired effect. Where differing amounts of echo are required on two voices taken on a single microphone the echo feed switch (changed between voices) should be used in preference to the echo output fader. Where the latter is varied it should be reset at the start of the new sound and not in the gap between sounds, where the change would be audible unless the gap were long.

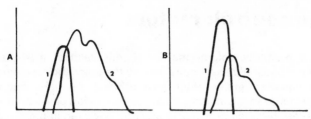

ECHO MIXTURES   With separate control of the direct feed, 1, and the echo feed, 2, a wide range of effects is possible. A, Reverberation dominates the sound. B, Reverberation tails gently behind it.

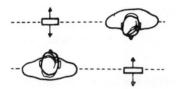

ECHO ON TWO VOICES   Where differing amounts of echo are needed on two voices opposing microphones and voices are placed dead-side-on to each other. This layout (using bi-directional microphones) might be adopted in a radio studio; in a television studio there will normally be more space, making separation easier.

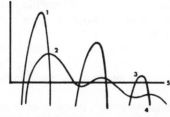

ECHO FOR PERSPECTIVE   A second bi-directional microphone, suspended above the main and having a stronger echo feed helps to create perspective on crowd scenes. Echo is automatically reduced as a speaker moves closer to the lower microphone.

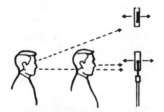

THE EFFECT OF LEVEL   1, Loud direct sound. 2, Audible reverberation. 3, Quiet direct sound. 4, Reverberation now below threshold of hearing, 5. For a given setting of the echo fader there is a level of direct sound below which the reverberation has no audible effect. This also varies with the setting of the listener's loudspeaker.

The assessment of relative sound volumes is done by ear,
subject to some reference level which is determined by meter.

Volume control: meters

The maximum permitted volume may be sharply defined. In AM radio '100 per cent modulation' means quite literally the point above which the peaks of the signal are greater than those of the carrier wave and will momentarily extinguish it. Similarly, on a record there is a physical limit to the wall thickness between grooves. In practice, however, in present-day terms 'full modulation' generally indicates an upper limit for some designated low level of irreversible distortion.

Similarly there is a lower limit to programme level below which noise of one sort or another becomes noticeable. Part of the sound man's job is to control and where necessary compress the signal to make the best use of the range between noise and distortion. Within the accepted range, he judges the relative levels by ear, but uses a meter to check the result.

Metering devices
The two main types of meter are the VU meter (used primarily in America) and the PPM (widely used in Europe).

The face of the VU meter has scales for both percentage modulation and decibels. It may be a direct-reading instrument – i.e. requiring no special amplifier. In a studio-quality meter the long time-constant (as much as 300 milliseconds is used to damp unreadable fluctuations) prevents transient sounds from being registered at all. It underreads on sharply percussive sounds and speech (at '100 per cent modulation' these are distorted).

A particular disadvantage of the scale is that 'percentage modulation' has little to do with relative levels as perceived by the ear, and that about half of the range is occupied by the 3dB above and below the nominal 100 per cent modulation. Very little natural sound varies over so narrow a range.

Nevertheless many users feel they get quite sufficient information from a VU meter and are happy with it.

The peak programme meter (PPM) is a programme aid as well as an engineering device. Using a special amplifier for conversion of the signal it reads logarithmically (i.e. in decibels) over a working range which is much wider than that of the VU meter. A typical PPM has a time constant of 2.5 milliseconds for the rise and a slow fall (time constant 1 second, giving a fall of 8.7dB/second).

Meters are also used to ensure that there is no uncalled for loss or gain between items of equipment; to control relative levels between different performances; and to check that levels remain within the bounds desirable for good listening.

In music recording studios, multitrack techniques ideally require separate monitoring for each channel that is being recorded. Using the conventional display, about eight monitors can be read. For a larger number it is best to have them in line with the fader and rising on a vertical scale to form a graphical display from left to right.

144

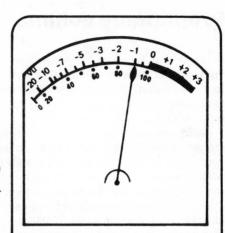

VU METER   The upper scale is measured in decibels relative to '100 per cent modulation'. Percentage modulation is indicated on the lower scale. Distortion is to be expected when the needle goes over the 100 mark for music or 50 for speech.

PPM (PEAK PROGRAMME METER)   The simple display (usually white on black) is easy to read over long periods. Figures on the PPM used by the BBC are arbitrarily chosen: between 2 and 6 each division represents 4dB; 6 represents 100 per cent modulation. The rapid rise reaches 80 per cent of the full deflection in 4 milliseconds: distortion on sounds which are of shorter duration than this are not noticeable. All meters within a given chain are lined up to a common reference level: they should all read '4' when a standard tone is fed through the system.

LED METER   A strip of light-emitting diodes responding instantaneously to volume. They are wearing to watch for long periods and are not used professionally.

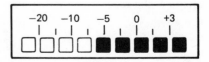

145

Programme volume

A meter can be used to suggest or check levels. For example, if a discussion is allowed its normal dynamic range with occasional peaks up to the maximum, a newsreader who speaks clearly and evenly will sound loud enough with peaks averaging about 6dB less. A whole system of such reference levels can be built up: see opposite.

Radio stations with a distinctive style and a restricted range of programme material have only a limited number of types of junction and it is easy to link item to item. But for a service (still available in many countries) which incorporates news, comedy, popular and serious music, religious services, magazines, discussion programmes and so on, problems of matching between successive items can become acute: a listener expects to adjust his volume only with increasing interest in a particular item. An increase in volume that coincides with a reduction of interest is irritating.

Junctions between speech and music

Two rules which generally smooth junctions between speech and music are given in the second table opposite. The apparent contradiction can be resolved by starting speech 2dB down from its normal level and then raising it by that small difference. These levels assume listening with at least moderate attention with the volume set accordingly. In these circumstances announcements should peak less than music. But if the listener's volume is always assumed to be low a higher relative level of speech is tolerable: in the worst conditions an announcement may have to be 8dB *above* music . . . if the announcer has anything to say which is actually worth hearing.

Problems of matching are of less interest to stations whose main audience may be listening in noisy or unfavourable conditions (e.g. in a car or on poor equipment such as a small transistor radio), or whose principal concern is maximum coverage – as in developing countries where a small number of transmitters cover a wide area and communication takes precedence over quality.

Preferred listening levels

An experiment on preferred listening levels (third table, opposite) shows that sound balancers prefer to listen at marginally higher levels than musicians; and both very much higher than most members of the public. People professionally concerned with sound need to extract a great deal more information from what they hear, and the greater volume helps them to pick up the finer points of fades and mixes, etc, and check technical quality. But it must always be remembered that the state of affairs at the receiving end may be very different.

146

RECOMMENDED PEAK LEVELS for a radio service with a full range of output. A restricted range of output may permit a narrower range of levels. Here, the levels given are relative to 100 per cent modulation at the transmitter. They were obtained by using a peak programme meter (PPM).

Programme material	Peak level, dB
Talk, discussion programmes	0
News and weather	−6
Drama: narration	−8
Drama:action	0 to −16
Light music	0 to −16
Serious music	0 to −22*
Harpsichords and bagpipes	−8
Clavichords and virginals	−16
Announcements between music (depending on type of music)	−4 to −8

* or lower for periods of up to half a minute

LISTENERS' PREFERENCES FOR RELATIVE LEVELS OF SPEECH AND MUSIC   Based on the results of a BBC survey. This included serious and light music (but not pop music) and was designed to measure preferences for listening in reasonable conditions at home. These are averaged results.

Speech following music to be 4dB down
Music following speech to be 2dB up

PREFERRED MAXIMUM SOUND LEVELS

| | Public | | Musicians | Sound-balancers | |
	men	women		men	women
Symphonic music	78	78	88	90	87
Light music	75	74	79	89	84
Dance music	75	73	79	89	83
Speech	71	71	74	84	77

These figures were obtained by BBC experiments in 1948, before the advent of either FM radio or popular music in its modern forms. But the trends shown here remain valid. (The figures are given in dB relative to 2×10^{-5} N/m².)

147

Manual control of music and speech

About 110dB separates the thresholds of hearing and of feeling. The significant sounds (including music) that we hear in our daily lives cover much of this range. But only 45dB may separate the noise level of a quiet room and its occupant's preferred maximum level of music. So the natural or possible range of sounds must be compressed to meet the preference of many listeners, as well as to hold levels between the noise and distortion levels of the recording media or transmission channels.

Controlling music

In the concert hall, music may have a range of 60–70dB. In BBC practice it is compressed to about 22dB (peak values) with quieter passages not exceeding half a minute. Records may be allowed more than this.

Automatic control without the anticipation of quiet or loud passages would destroy the music, but manual control can enhance the listener's enjoyment. Avoid overmodulation by fading down gradually over a period of half a minute or more; anticipate very quiet passages similarly. Try to preserve the nuances of light and shade. Steps of 1½–2dB at a time will not be noticed by the audience; most faders are marked in 'stops' of about this size.

Controlling speech and sound effects

Speech is controlled in a different way and over a narrower range. The maximum range is in radio plays, where BBC practice allows 16dB (again, on peak values), with the average at 8dB below the maximum to allow for the dramatic effect of louder passages. Even so, shouting has to be held back (making it sound distant) unless the performers restrict their own volume and project the effect of shouting. Manual control of levels may be exercised more abruptly than for music if the adjustment is made at the end of the gap between one voice and another. A sudden adjustment by as much as 6dB made during the first syllable following an unexpected change in voice level may be barely noticeable.

The *volume* of a percussive sound effect such as a pistol shot or car crash must be suggested by the *character* of continuing sound. Fortunately this necessary convention is accepted by listeners.

Recording for film

In recording for film, the minimum of control should be exercised at source. The purpose of this is not to match the voices in different shots: these can be put on separate tracks and matched at the dubbing (re-recording) stage. Rather, it will help the editor with the background atmosphere – for example, tiny gaps that are opened out can be filled by cutting in sound trimmed from other parts of the same take.

148

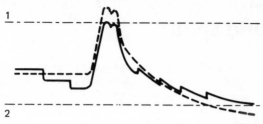

MANUAL CONTROL OF PROGRAMME VOLUME   Method of compressing a heavy peak to retain the original dramatic effect and at the same time ensure that the signal falls between maximum and minimum permissible levels 1 and 2. Steps of 1½–2dB are made at intervals of, say, ten seconds.

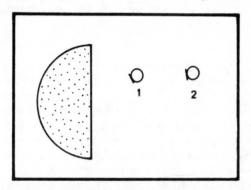

VOLUME CONTROL AND PERSPECTIVE   Fading up a quiet passage is like moving closer. But if this is done on a single microphone the perspective remains the same. A pleasanter effect may be achieved by mixing microphones at two distances: microphone 1 is used to bring forward the quiet passages and microphone 2 takes over in the loudest parts of the music.

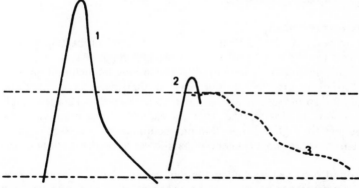

CHANGE IN DYNAMICS OF A SOUND EFFECT   An effect of which a dominant element is sheer volume of sound, such as a rifle shot, 1, must be held down close to the 100 per cent modulation level, 2 (though some distortion due to overmodulation is acceptable). The loudness of the sound must be suggested by the echoing reverberation that follows it, 3. Additional sounds may be used to confirm its nature: the classical example is the ricochet.

Monitoring sound

By monitoring we mean listening to sound quality, operational technique and programme content at the point of origin. Further along the line it is used to check for changes in quality due to losses in recording, transmission by land-line, and so on.

To make effective use of his aural judgment the balancer needs good listening conditions and a high-quality loudspeaker. So the main thing that distinguishes a studio from any other place where a microphone and recorder may be set up is that it includes at least two acoustically separate rooms: one for the sound source and microphones, and the other where these are monitored and controlled.

Checking recordings

Because it is no good finding a technical fault ten minutes after the artist has left the studio, a recording should always be monitored at the time that it is made or immediately afterwards. As tape recorders necessarily have a space between recording and replay heads, there is a delay of a fraction of a second in sound monitored from tape. It is not easy to control sound which is monitored from replay: a fade may have to be controlled to an accuracy of a twentieth of a second, whereas the tape delay may be a fifth of a second or more. A complex recording should be monitored by a second individual (the recordist) and again by loudspeaker in his own acoustically separate area.

A monitoring loudspeaker for monophonic sound should be at about 3–6ft from the listener. At less than this the sound is modified by standing waves near the cabinet; at greater distances any deficiencies in the listening room acoustics take over.

The line-up procedure

To line up equipment to a common standard, an electrically-generated tone is fed from the control desk or mixer through the chain to the recorder or transmitter. A sample of this tone may be included on each tape recorded for professional purposes, so that replay equipment and subsequent recordings may also be lined up to the original standard. It is also common practice to take 'level', a sample of the material to be recorded. Fed through the chain (but not recorded) this serves as a test of the system in more practical conditions and confirms the identity of the source.

The sound recordist (including the film sound man) on location may have to simplify these procedures: he will be balancing sound in difficult conditions, controlling, mixing and monitoring replay on headphones all at the same time. It is a measure of his skill that even so he often achieves reasonably good quality.

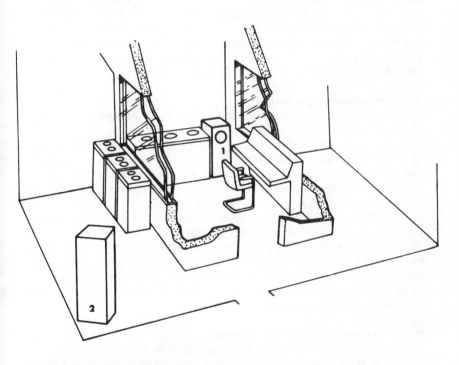

STUDIO, CONTROL AND RECORDING ROOMS   Ideally these are
acoustically separate, with high-quality loudspeakers, 1 and 2, at each point
where sound must be monitored. The desk has a tone source or generator,
the output from which can be checked by meter in the recording room and
recorded on tape for subsequent checks. The BBC uses 1000Hz at a level of 1
milliwatt in 600ohms as a standard throughout its equipment. This reads 4 on
a BBC peak programme meter and represents 40 per cent modulation at the
transmitter.

151

Listening for sound quality

Sound is monitored for deficiencies of various types:

1. Production faults such as miscast voices, stilted speaking of the lines, uncolloquial scripts, bad timing.
2. Faulty techniques, for example poor balance and control, untidy fades and mixes, misuse of acoustics.
3. Poor sound quality, as in distortion, lack of bass or top, or other irregularities in the frequency response.
4. Equipment or recording faults such as wow, flutter, hum, tape hiss, noises due to loose connections or poor screening.

Of these, the faults in the first group are certainly the most important. They are also the easiest to spot. And as for techniques: the ability to see faults here comes with the practice of the techniques. But the various things that can be wrong with the quality of a sound – objective, measurable things – are, surprisingly, among the most difficult to judge in other than subjective terms. The quality of sound that the ear will accept – and prefer – depends almost entirely on what the listener is used to. Few people whose ears are untrained can judge quality objectively; most people prefer the familiar medium to the unfamiliar.

Types of sound degradation

It is beyond the scope of this book to discuss quality and equipment faults in detail, but here are a few examples:

A tape-recording head that has become permanently magnetised over-records on one half of the sound cycle, so that distortion sets in at a lower level. If recording and replay heads are set at a different angle there is high-frequency loss. If part of the coating is missing, dropouts occur. In storage, tape may be subject to printing of the signal from one layer to the next, or the tape itself may become stretched. The tape drive spindle may be eccentric, causing flutter. A record-player motor may cause rumble. A disc may be warped or the hole or the turntable itself may be off-centre, causing wow. Distortion on disc may be caused by a damaged groove or worn stylus, and so on. And the distortion of electrical signals and the induction of spurious signals such as studio talkback or mains hum can arise in many different ways.

Keeping the problems in proportion

Audio engineers naturally have a strong interest in these matters and may comment exhaustively on faults that are small in comparison with other problems that the balancer faces. A good sound man should be able to see all of these in their proper proportions: he must be able to solve both technical and aesthetic problems together.

HIGH-FREQUENCY LOSSES

A, Progressive loss due, for example, to transmission by landline. Top can be restored, but noise will rise with it. B, Extinction at low frequencies, due for example to phase cancellation or azimuth misalignment. In this case, top cannot be fully restored.

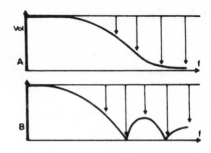

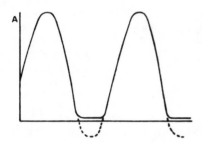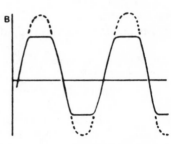

HARMONIC DISTORTION

The dotted portion of the original signal is flattened. A, Bottoming causes each alternate peak to be clipped. B, Both peaks are flattened, as excessive swings in each direction cannot be followed by some item of equipment. In both cases the distortion produced includes tones which are exact multiples of the frequency of clipping. Because the new frequencies are normally present in musical sound, the ear is tolerant of this form of distortion: 2 per cent third harmonics will be barely noticeable on much sound, and 1 per cent is generally acceptable in equipment specifications. *Intermodulation distortion,* in which sum and difference tones are produced, is much less tolerable as these are not naturally present in sound.

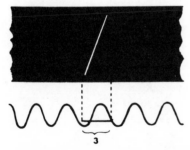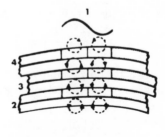

LOSS OF QUALITY

There are many possible mechanisms that are unrelated to deficiencies of microphones or acoustics but which introduce similar effects. Two examples are: *Left,* azimuth misalignment of a tape head. 1, Tape. 2, Tape head at an angle (this ought to be perpendicular to the line of travel). 3, Extinction wavelength. *Right,* tape print through producing spurious echoes (including 'pre-echo'). 1, Wavelength of original and printed signal. 3, Layer with original signal. 2 and 4, Adjacent layers with printed signals.

153

Useful formulae

An understanding of these relationships is not essential to those with a purely practical interest in microphone use. But they will be useful to those with a mathematical or engineering background.

Frequency, wavelength and the speed of sound

$$f\lambda = c$$

where λ is wavelength
   c is the speed of sound
   f is frequency in hertz (Hz)
   1Hz is one cycle per second; 1kHz is 1000Hz.

The speed of sound varies with temperature:

$$c = 1087 + 2T \text{ ft/sec}$$

where T is the temperature in degrees centigrade

The speed of sound also varies with the nature of the medium. In fully saturated damp air it is some 3ft/sec faster than in dry air. In liquids and solids it is much faster.

Volume, intensity and power

$$\text{Volume} = 10 \log \frac{I_2}{I_1} \text{ decibels}$$

where I_1 and I_2 are the intensities.

The range of sound intensities between the threshold of hearing and the threshold of feeling (if this is accepted to be 120dB) is 10^{12} or 1 000 000 000 000:1.

The gain of an amplifier is also measured in decibels:

$$\text{Gain} = 10 \log \frac{P_2}{P_1} \text{ decibels}$$

where P_1 is the power input and P_2 the power output.
But power is proportional to the square of voltage, so

$$\text{Gain} = 10 \log \left(\frac{V_2}{V_1} \right)^2 = 20 \log \frac{V_2}{V_1} \text{ decibels}$$

A range of 1 000 000 000 000:1 in intensity is equivalent to only 1 000 000:1 in voltage.

Note that the voltage scale is used for microphones: dBs of voltage gain are not the same as dBs of increase in sound intensity.

154

Further reading

Nisbett, Alec. The Technique of the Sound Studio. *Focal Press,* London and Boston (4th Ed., 1979).
   Extends beyond the field of *The Use of Microphones* to include a full account of stereo techniques, sound effects, programme construction (including the use of disc and tape, the function of fades and mixes, etc.), radiophonic techniques (including musique concrète and electronic music), tape editing, and the role of sound in television and film.

Millerson, Gerald. The Technique of Television Production. *Focal Press,* London and *Hastings House,* New York. (10th Ed., 1979).
   A highly analytical study of the medium, including the contribution of sound.

Oringel, Robert. Audio Control Handbook, *Hastings House,* New York. (4th Ed., 1972).
   An elementary introduction to the subject (a little fuller in this edition than in the first three). There is a strong emphasis on American equipment.

Wood, A.B. The Physics of Music, *Methuen,* London. (7th Ed., 1975).
   A standard primer on the subject.

Potter, Kopp and Green-Kopp. Visible Speech. Constable, London. (1966). Dover Publications, New York.
   Provides a clear understanding of the characteristics of the human voice.

Hilliard, Robert L. (Ed.). Radio Broadcasting. *Hastings House,* New York. (2nd Ed., 1974).
   An 'introduction to sound medium', including American production, programming and directing methods.

Robertson, A.E. Microphones. *Iliffe,* London. (2nd Ed., 1963).
   Describes the engineering principles behind many types of microphones. Technically complex.

Johnson, Joseph S. and Jones, Kenneth K. Modern Radio Station Practices. *Wadsworth,* Belmont, California. (1972) and *Prentice Hall,* Hemel Hempstead.
   The aims, motives and production methods of American radio.

Glossary

Note on differences in British and American usage

This book is designed to be read in both America and Britain, as well as many other parts of the world. The reader is asked to forgive the mid-Atlantic terminology that often results. Differences of usage appear not only between countries but even from one organisation to another: both BBC and NBC have their own 'house' terminology. Further anomalies may appear in translation from other languages, or by the whim of the engineers or salesmen for manufacturing companies in the description of their products.

There are variations in ways of describing the job this book is about, the place where it is carried out, and the equipment used.

The job itself

The American term audio operator broadly describes the job. But I betray the English origins of this book by avoiding the term in the text: 'audio' and 'video' are used much more in America than in Britain where the older 'sound' and 'picture' are preferred except in certain compound terms of American origin or for differentiating between signal feeds in television recording systems. In BBC radio an audio operator is called a studio manager. Television 'sound supervisors' and film 'sound recordists' are not unreasonable names – although purists may object to 'recordist' as a way of avoiding confusion between the man and his machine, the recorder. In this book I have avoided most of these by simply calling the sound man a sound man, or if his function warrants it at that moment, the sound balancer.

The place and the equipment

The place where much of the job is done might most reasonably be called the control room or, in television, sound or audio control. BBC radio is bound by its own history: there 'control room' was the name given to the main switching centre, and the room that is part of a studio suite is called a control cubicle. The name was originally accurate: it started life as a structure like an overgrown telephone box in the corner of the studio.

The control desk is also called the panel (BBC radio) or the board (US usage). The master control (US) is also called a main gain control, and a sub-master control is a group fader. A British outside broadcast has Americans puzzling: it turns out to be a remote; while a nemo on the NBC board might confuse a British sound man, who would call it an outside source. Readers may add their own examples from these pages.

Glossary

Absorption (30–31) (75–77) Loss of sound energy due to friction as the air moves back and forth in the interstices of a porous material (a *soft absorber*) or by transfer to another vibrating system material from which less sound is subsequently re-radiated. See also *membrane absorber, Helmholtz resonator.*

Absorption coefficient (30) The fraction of sound energy which is absorbed, usually on reflection at a surface and unless otherwise stated for a frequency of 512Hz at normal incidence.

Acoustics (28–30) (38) (58) (66–69) (70) (74–77) (86–88) (110) (124) (138) The behaviour of sound, and its study. The acoustics of a studio depend on its size and shape and the amount and position of absorbing and reflecting materials.

Amplification (9) (44–45) (55) (120) (128) (154) A gain in strength, usually of an electrical signal.

Antinode (22–23) The part of a stationary wave where there is maximum displacement.

Atmosphere (72) (148) The normal background sound at any location.

Attack (20) The way a sound starts. See also *transient.*

Attenuation (104) (131) (137) Fixed or variable losses, usually of an electrical signal.

Audio frequencies (12) Those which lie within the range of good human hearing: say 16Hz–16000Hz. An *audio signal* is an electrical or electro-magnetic transmission containing audio frequency information.

Backing track (98) A pre-recorded musical accompaniment to which the soloist can listen on headphones as he adds his own performance.

Baffle (96) A small acoustic screen which causes a local variation in the acoustic field near a microphone. It distorts the high-frequency sound, by setting up a system of standing waves.

Balance (8) (66–77) (86–127) The choice of microphone positions to pick up an adequate signal, discriminate against noise and provide an appropriate ratio of direct to indirect sound. It may also include the creative modification of the electrical signal.

Balance test (67) (86) (94) A trial balance, or a series of trial balances which should preferably be judged by direct comparison.

Bass (40) (50) (68) Lower end of the musical scale. In acoustics, the range (below about 200Hz) in which there are difficulties, principally in the reproduction of sound, due to the large wavelengths involved.

Bass tip-up (40–42) (46) (52–53) (68) (91) (92) (108) A selective emphasis of bass which occurs when a microphone responding to pressure gradient is placed where there is a substantial reduction in sound intensity between the two points at which the sound wave is sampled. It is most noticeable when the microphone is close to the source.

Baton microphone (56) Hand-held stick microphone.

Bi-directional microphone (34–35) (38–46) (68–71) (74–77) (87) (93) (99) (103) (143) One which responds to sound from its front and rear but not to sound from its sides and above and below.

Boom (42) (50) (65) (79–85) An arm, usually telescopic and mounted on a floor stand or dolly from which a microphone is slung.

Buzz (72) Colloquial term for location sound atmosphere (q.v.).

Cable (54) (58) (60) (82) Generally the free wiring between microphone, etc., and fixed plug-in point.

Cancellation (14) (36–39) (48–50) (57) Partial or complete opposition in phase, so that the sum of two signals approaches or reaches zero.

Capsule (55) A plug-in microphone component containing the diaphragm, but not the head amplifier or transformer.

Cardioid response (34) (42–46) (50) (57) (68–71) (74) (80) Literally, a heart-shaped directional response.

Cavity resonance (16) (30) (31) (36) (58) See *Helmholtz resonator.*

Clean feed (128) (131) A cue feed to a remote programme source which includes all but the contribution from that source.

Coloration (28) (68) (86) (138) Distortion of frequency response by resonances at particular frequencies.

Compression (60) (86) (128) (136) (149) Control of sound levels to ensure that all wanted signals are suitably placed between the noise and distortion levels of the medium, and that relative levels are acceptable to the intended audience. *Manual and automatic compression* are used for different purposes.

Compression ratio (136–137) A selected degree of automatic compression, e.g. 2:1, 3:1, or 5:1. It is set to operate at a given onset volume, e.g. 8dB below full modulation.

Condenser microphone (32–34) (42–46) (55–59) (62) This depends on the electrical quality of capacitance, i.e. the ability of neighbouring and oppositely charged conductors to store energy between them. A condenser is a device for doing this, and in a condenser microphone the two components are two diaphragms or a diaphragm and base plate. Variations in their distance apart produce corresponding fluctuations in capacitance and so generate an electrical signal.

Contact microphone (120) (141) One which directly picks up the sound transmitted by a solid material.

Control (52) (128–152) The adjustment of programme level (in the form of an electrical signal) to make it suitable for feeding to a recorder or transmitter.

Cottage-loaf response (34) (46–47) A bi-directional response in which pick up at the front is more sensitive and covers a wider angle than that at the rear. See *supercardioid response.*

Crystal microphone (32–34) This employs slices of certain crystal or ceramic materials which are cut and placed together in a particular way. The bi-morph so formed produces a voltage between opposite faces when the crystals are made to bend together.

158

Dead room (76–77) A room with very thick sound absorbers (often 1 metre deep).

Decay (138) The way in which a note ends; or in which a sound or its reverberation dies away.

Decibel, dB (26) (144) (147) (154) A measure of relative intensity, power or voltage.

Delay (140) (142) Storing a sound for a moment, usually on tape or in a digital store, then recombining it with the original signal.

Diaphragm (10–12) (33–37) (42–44) (54) (56) That part of the microphone which responds to air pressure – or pressure gradient, if exposed on both sides.

Diffusion (28) (30) Breaking up sound waves by means of irregularly distributed reflecting surfaces.

Digital audio signal (130) (134) (140–141) Audio information which has been converted from *analogue* form (in which the electrical waveform is directly related to the sound waveform) to a binary code. While in this form it is not subject to the gradual accretion of noise which continuously degrades analogue signals.

Digital echo (140–141) Reverberation generated by using a digital signal and random access memories.

Directivity (47) The front to back ratio of sound picked up by a microphone.

Distortion (9) (32) (66) (148) (152–153) Unwanted changes of sound quality, generally by the introduction of electrically generated tones, or by changes in the relative levels of the different frequencies present.

Dynamic microphone (32) Imprecise term usually meaning moving coil microphone.

Dynamics (137) (149) The way in which sound volume varies.

Echo (9) (114) (128) (131) (138–143) Literally, the discrete repetition of a sound at least a twentieth of a second later; colloquially, artificial reverberation.

Echo chamber (138–140) Reverberant room, through which a signal is fed via loudspeaker and microphone.

Echo plate (62) (140–141) A sheet of metal which is used to provide artificial reverberation. It is vibrated by one transducer; another responds to the reflected wave motions.

Eigentone (28) A characteristic resonance due to the formation of standing waves between parallel walls.

Electret (33) (44) (58–59) Miniature, lightweight electrostatic microphone capsule.

Electromagnetic wave (8) Radiant energy travelling at the speed of light. This includes gamma and X-rays, light, radiant heat, and radio.

Electrostatic microphone (44) See *condenser*.

End-fire microphone (42) (54) One with the main axis of polar response in line with the axis of the body of the microphone. The diaphragm is in a plane at right-angles to this axis.

Envelope (20) The way in which the volume of a sound varies in time; its dynamics (q.v.).

Equalisation (9) (56) (86) (120) (128) (132) (135) Changes in the electrical signal, nominally to correct for frequency distortion introduced at any stage in an audio system. In practice it is also used for creatively distorting the frequency response still further, the result being judged by ear.

Exponential A rate of growth in which doubling occurs at equal intervals. See *logarithmic scale*.

Fade (128–131) (148) Gradual reduction or increase in an electrical audio signal by the use of a fader or 'pot' (potentiometer).

Field pattern (34) Polar response (q.v.).

Figure-eight response (35) (38–46) Also called 'figure-of-eight'. A bi-directional microphone response in which the front and back are live but opposite in phase. See also *bi-directional*.

Filter (40) (44) (128) (132–135) A network of electronic elements (generally resistances and capacitances) which allows some frequencies to pass and attenuates others.

Fishpole (83) A hand-held microphone boom.

Flutter (152) Rapid fluctuation in pitch, generally of 8Hz or more. It is often due to a mechanical equipment fault such as an eccentric tape drive spindle. See also *wow*, *vibrate*.

Flutter echo (28) A rapid fluctuation in volume which is heard when a staccato sound reverberates between parallel reflecting walls.

Foldback (128–131) (139) A feed of selected sources to a studio loudspeaker for the benefit of the performers.

Formant (20–25) (134) A broad frequency resonance associated with a sound source, e.g. a voice or a musical instrument. It is caused by the physical characteristics (notably the dimensions) of the resonant system.

Frequency (12–19) (37) (154) In pure tones, the number of complete oscillations that an air particle makes about a median position in one second. Complex sounds are made up by adding many such simple patterns of motion (or frequencies).

Frequency response (32) (36–61) (132) The way in which the relative levels of the different frequencies in a sound or audio signal are changed in passing through a stage or electrical component in its path (e.g. by a microphone). *Frequency correction* may be introduced to restore or otherwise deliberately change the frequency response. This is also called *equalisation*.

Fundamental (13) (18–25) The primary and usually the lowest component of a musical note, and that which defines its pitch.

Gain (154) Amplification, generally calculated in decibels.

Gun microphone (48–51) (63) (70) (80) (115) Microphone fitted with interference tube to make it highly directional.

Harmonics (18–21) (25) (153) A series of frequencies that are all multiples of the lowest frequency present (the fundamental). *Harmonic distortion* is the production of unwanted harmonics, e.g. by flattening the peaks of waveforms that are too big for the electrical response of the equipment they are passing through.

Head amplifier (44–45) (55) A small amplifier within the microphone casing which converts the fluctuating capacitance of a condenser microphone into the form of an alternating current which is suitable for transmission by cable to a mixer or recorder.

Helmholtz resonator (16) (30–31) A cavity within which air can be made to expand and contract at a frequency which is characteristic of its size and shape. If a soft absorbing material is placed in the mouth of the cavity, it will selectively reduce sound at the resonant frequency.

Hertz, Hz (12) (154) Number of cycles or excursions per second in a pure tone.

Howl-round or **howl-back** (88) (140) Closed circuit, e.g. microphone, amplifier, loudspeaker and sound path back to the microphone again, in which the overall gain exceeds the losses of the system.

Humidity (88) (110) (138) Damp air transmits sound better than dry air, in which upper harmonics are attenuated in a distant balance. A close balance therefore sounds more brilliant than any performance heard at normal distance in a concert hall.

Hypercardioid response (46) Imprecise manufacturers' term; probably the same as supercardioid or cottage-loaf – but this should be confirmed by looking at a polar diagram.

Impedance (32) (64) The combination of resistance, inductance and capacitance that serves to reduce the signal in an alternating current. An inductance selectively reduces high frequencies; a capacitance discriminates against the low; a resistance acts equally on all frequencies.

Intensity of sound (16–17) (26) (40) (154) The sound energy crossing one square metre.

Interference tube (50–51) (57) An acoustic channel which, attached to a microphone capsule, confers highly directional properties. Sounds approaching from an angle to the axis are reduced or eliminated by phase cancellation.

Intermodulation distortion (153) Unwanted audiofrequencies that are generated as the sum and difference of frequencies present in the original signal. It is much more noticeable than harmonic distortion (which is often masked by the harmonics in naturally occurring musical sound).

Lavalier (58–59) Microphone suspended round the neck. Also called neck or lanyard microphone.

Level (137) (143–149) Volume of the electrical signal. A *level test* is used to find suitable settings for (a) the source fader, to allow for expected variations in sound volume, and (b) the master control, to feed the signal onward at a level which is appropriately placed between the noise and distortion levels of the following equipment.

Limiter (60) (136) An automatic control to reduce volume when the signal rises above a level that would introduce significant distortion.

Line up (150) To arrange that the signal passes through all items of equipment at an appropriate *level* (q.v.). This is achieved by the use of *line-up tone,* often at 1000Hz, which is fed through the entire system and should produce a standard reading on meters at every stage.

Lip-ribbon microphone (52–53) One which is held at a close, standard distance from the mouth. Equalised bass tip-up gives added discrimination against low frequencies.

Location (72) Place outside a studio used to record or film sound and/or picture.

Logarithmic scale (18) (26–27) (144) (154) One which converts a particular form of growth ('exponential', q.v.) back to a linear scale. Differences in sound intensity or frequency which increase in the ratios 1:2:4:8:16:32:64 appear to the ear to grow by equal intervals, so it is convenient to represent them by a scale which also has equal intervals – 0:1:2:3:4:5:6. This is a logarithmic scale.

Loudness (26–27) (146–148). The subjective aspect of sound intensity.

Membrane absorber (30) (31) A damped resonating panel which will respond to and selectively absorb a range of sound frequencies.

Meter (9) (129) (132) (144–145) Device for measuring voltage, current etc. The *VU meter* and *peak programme meter* (q.v.) are adaptations for measuring audio-signal volume.

Microphone (8) (24) (32–63) A device for converting sound to electrical energy.

Midlift (102) (104) (108–109) (120) (134) Deliberately-introduced peak in the middle- and upper-frequency response.

Mix (9) (70) (128) The electrical combination of audio signals. In a *mixer* each source is first controlled by its own fader.

Modulation (136) (144) (148) Superimposition of the audio signal on a carrier wave of higher frequency. *100 per cent modulation* or *full modulation* corresponds to the maximum acceptable audio signal level beyond which overload distortion occurs.

Monitor (8–9) (111) (140) (150–152) Check sound quality by ear. A *monitoring loudspeaker* is used to check the aesthetic quality as well. (In addition, picture monitors are used in television, both in the control room and on the studio floor).

Mono, monophonic sound (8) (28) Sound combined to be reproduced through a single loudspeaker. If several loudspeakers are used, the signal to each is the same.

Mouse (64–65) (116) Microphone in soft, acoustically transparent housing attached to solid, reflective surface.

Moving coil microphone (32–34) (42–43) (56) (58) (62) One in which a coil attached to the diaphragm moves in the field of a magnet.

Music balance (86–127) See *balance. For individual instruments see contents list.*

Newtons per square metre, N/m² (26) (32) (147) The unit of sound pressure. This has replaced an older measure, dynes per square centimetre. $1 N/m^2 = 10 dynes/cm^2$. A common reference level, $2 \times 10^{-5} N/m^2$ approximates to the threshold of hearing at 1000Hz.

Node (19) (22) In a stationary wave, a point at which there is no displacement.

Noise (9) (32) (50–52) (66) (72–73) (76) (100) (128) (130) (136) (148) (152) Unwanted sound or audio signal.

Noise reduction system (130) This divides the audio-frequency spectrum into several bands: each is then separately and automatically lifted to its optimum signal-to-noise level before recording, and is reduced accordingly on replay. This permits multiple re-recording without the introduction of appreciable noise from the tape itself. (It is generally used to minimise the noise from recording systems, but in principle can be used around any noisy element in a sound system.) The use of digital signals serves a similar purpose.

Obstacle effect (37) (103) A sound is not impeded by obstacles which are smaller than its own wavelength but is reflected or absorbed by those which are larger.

Omnidirectional response (34) (37) (56–59) (62) (68) Responding equally to sound from all directions.

Outside source (128) British term for remote source.

Overtones (18–20) Individual component frequencies in a sound which, when added to the fundamental, help to define its musical quality.

PA, public address (128) (131) A feed of selected sound sources to audience loudspeakers.

Parabolic reflector (48–49) A reflector which will concentrate sound from a distance and in a given direction to a single point. It is efficient only at wavelengths less than its diameter.

Peak programme meter, PPM (144–147) A meter used in Britain for measuring the peak values of programme volume.

Phase (13) (14) (34) (38–42) (48) (57) The stage which a particle responding to a pure tone has reached in vibrating about its median position. Particles which are at the same stage in this cycle of movement are said to be *in phase.*

Phase cancellation (38) (48–49) (57) The superimposition of two waves where one is positive and the other negative, so that their total is less than either on its own.

Phasing (140–141) An effect in which two similar or identical signals are very slightly separated in time, so that the waveforms are enhanced at some frequencies and cancel at others. If the time separation is changed, a hollow, sweeping effect is heard. Also called flanging.

163

Phon (26) A unit of subjective loudness. Phons equal decibels (measured objectively) at 1000Hz, and at other frequencies are related to them by contours of equal loudness.

Piezo-electric effect (33) The operating principle of a crystal microphone (q.v.).

Pitch (18) (22) The subjective aspect of frequency. In harmonic series, the frequency of the fundamental.

Polar characteristic, polar diagram, polar response (34–59) (88) These describe the way in which the response of a microphone varies with the angle of incidence of a sound. Since this will change with frequency, it is given for a representative series of pure tones. For each of these the response along the main axis of symmetry (generally a line normal to the diaphragm) is taken as unity, and the response at other angles is related to this.

Post-balancing (130) Treating (by equalisation, etc.) and mixing sound that has already been recorded on multitrack tape.

Pre-amplifier (9) (128) Amplifier before first fader.

Pre-hear (128) A means of sampling a programme source before fading it up and mixing it in to the studio sound.

Presence (56) (68) (102) (109) (134) Bringing 'forward' an instrument (or voice) by selectively amplifying a range of frequencies which contains much of its character.

Pressure gradient (14) (34) (40) (44) The difference between two successive points in a sound wave. A microphone which measures pressure gradient (on opposite sides of the diaphragm) will have a directional response, whereas one which measures the pressure at a single point will be omnidirectional.

Progressive wave (22) A wave which travels through a medium (as distinct from a standing or stationary wave).

Pure tone (10) (150) A sound or signal containing one audio frequency only. See also *sine wave, line-up tone.*

Radio microphone (60) Microphone attached to a small radio transmitter, with a receiver to link it to sound control.

Reference level (26) (147) (150) Since decibels are a measure based on ratios there is no absolute zero, and levels are given relative to some arbitrary reference level. For sound a useful reference level is 2×10^{-5} N/m^2 as this roughly corresponds to the lower limit of human hearing at 1000Hz. See also *line-up tone.*

Resonance (16) (30) (36) (90) (138) (140) Natural periodicity; reinforcement associated with this.

Response (32–59) (132–135) Sensitivity, frequency and polar characteristics of a microphone.

Response selection (132–135) The corrective or creative manipulation of the frequency content of a signal.

Reverberation (9) (28) (38) (46) (50) (68) (76) (78) (86–87) (116) (138–142) The sum of many reflections of a sound in an enclosed space.

Reverberation time (28–29) (74) (88) (138) (142) The time it takes for a sound to die through 60dB.

Ribbon microphone (32–34) (38–39) (42) (46) (52) (75) One in which a narrow strip of foil is suspended in a magnetic field. A fluctuating current is produced by movement of the foil with air pressure – or, more commonly, pressure gradient.

Scale (13) (18–19) Division of the audio frequency range by musical intervals, i.e. frequency ratios. The more harmonious intervals have simple frequency ratios, e.g. 1:2 (octave), 2:3 (fifth), 3:4 (fourth), etc.

Screen (30) (74–77) (87) (107) (109) (120–126) A free-standing sound-reflecting or absorbing panel used for the local modification of studio acoustics.

Sensitivity (10) (32) (56) Microphone output measured in decibels relative to one volt per Newton per square metre.

Separation (104) (121) (122) The degree to which the signals from different sound sources are kept apart for the purpose of individual treatment and control.

Shotgun microphone (50) See *gun microphone*.

Sibilance (24) (68) The over-emphasis of 's' and 'ch' sounds in speech.

Signal (8) (66) (136) The required information content of a sound field or of an electrical or electromagnetic transmission.

Signal-to-noise ratio (9) (32) The ratio of information content to unwanted hiss, rumble, hum and other unwanted background noises, measured in decibels.

Sine wave, sine tone (10) A wave containing a single frequency; a pure tone (q.v.). In mathematics, 'sine function' describes the shape of this simplest type of wave. The term is avoided in this text.

Sound (10–31) A series of compressions and rarefactions travelling through air or another medium, caused by some body or bodies in vibration.

Speech balance (24) (68–77) (116) (148) See *balance*.

Spill (118) (122) (132) Sound picked up by a microphone other than that intended.

Spin (140) An effect obtained by recirculating an original signal repeatedly through the same system. Any imbalance in the frequency characteristics of the system is rapidly enhanced. See also *howl-round*.

Spotting (110) Setting additional microphone to selectively enhance some part (usually an instrument or group) within an overall balance.

Springs (138) Reverberation-producing device in which a signal is sent along metal arranged in a coil, to be reflected many times at discontinuities in the material of which it is made. The springiness of the coil is not actually used.

Stand (54) (62–63) (71) (100) Microphone mounting.

Standing wave, stationary wave (22) (36) (75) The sum of two equal waves travelling in opposite directions. This may be caused by reflection at a wall or at the end of a pipe.

Stereo, stereophonic sound (8) (36) (46) (98) (110) (116) (129) Sound combined to be reproduced through two or more loudspeakers, each with a different signal, in order to give an impression of spatial spread.

Studio (sound studio) (28–30) (75) (79) (151) An enclosed space designed or primarily used for microphone work.

Supercardioid response (34–35) (46) (69) (92) (102–103) (108) (120) Unclear term for the much more descriptive 'cottage-loaf' response. In this book it is taken to be a polar response intermediate between cardioid and bi-directional. See *directivity*.

Talk-back (82) (152) A microphone in the control room linked to a loudspeaker or headphones in the studio.

Tone (14) (150) In audio practice, a single-frequency signal or sound. See *sine wave*, *zero level tone*.

Tone control (132–133) Alteration of bass and treble ('top') content of audio signal. See also *equalisation*.

Tracking (130) Building up a recording in successive stages. Later tracks are recorded to the replay of those already on tape.

Transducer (140–141) A device for converting audio information from one medium (sound, electrical current, electro-magnetic transmission, disc, tape, etc) to another.

Transformer (33) (129) (139) A device which isolates the direct current components of an electrical circuit while permitting the signal to pass with no significant loss of power, but changing voltage and impedance as required. Many microphones produce an electrical signal which is unsuitable for feeding along long cables or into subsequent equipment; the transformer provides the necessary matching.

Transient (86) (94–95) (102) In this context, often the irregular initial part of a sound before a regular waveform is established; an important part of the character of musical instruments.

Unidirectional response (34) (70) Nominally, a microphone which is live on one side only. In practice this is not achieved with much fidelity. The most strongly single-sided response is a particular kind of supercardioid. The term is also more often applied to a cardioid response (in which the whole of one side picks up a good-quality signal) or to a bi-directional response in which the back component is degraded or shielded, rather than to the highly directional response of, for example, a gun microphone. See *directivity*.

Velocity of sound (10) (12–13) (154) At room temperature this is approximately 1120ft/sec, and for most practical calculations may be regarded as constant. In liquids and solids it is slower.

Velocity microphone (32) Imprecise term usually meaning ribbon microphone.

Vibrato Rapid cyclic variation in pitch at about 5–8Hz. See also *wow*, *flutter*.

Volume (26) (40) (144–149) (154) See *level*.

VU meter (144–145) Volume unit meter, used in much American equipment for measuring audio signal levels. It shows percentage modulation and is not linear in decibels.

Wave, sound wave (9–17) (22) (28) (32) (48–50) A succession of compressions and rarefactions transmitted through a medium at a constant velocity, the speed of sound.

Wavelength (10–132) (48–50) (154) Provided that the wave is perfectly regular (i.e. a sine wave or pure tone) this is the distance between successive peaks.

Windshield, windscreen (32) (42) (50–51) (53) (64–65) Shield which fits over a microphone to contour it for smoother airflow, thereby reducing or eliminating wind noise.

Wireless microphone (60) See *radio microphone*.

Wow (152) Cyclic fluctuation in pitch due to mechanical variation in speed in a recording, with a frequency below about 5Hz. See also *flutter, vibrato*.

Zero level tone (150) Electrically generated tone at a standard reference level used to line up electrical equipment. In BBC practice this is one milliwatt in 600ohms using 1000Hz pure tone.